THE SPLENDID VOYAGE

THE SPLENDID VOYAGE

AN INTRODUCTION TO NEW SCIENCES AND NEW TECHNOLOGIES

Pangratios Papacosta

FRONTIERS OF SCIENCE

Prentice Hall Press • New York

Published by Prentice Hall Press
A Division of Simon & Schuster, Inc.
Gulf+Western Building
One Gulf+Western Plaza
New York, NY 10023

PRENTICE HALL PRESS is a trademark of Simon & Schuster, Inc.

Library of Congress Cataloging-in-Publication Data

Papacosta, Pangratios.
The splendid voyage.

(Frontiers of science)
Bibliography: p.
1. Science. 2. Technology. I. Title. II. Series.
Q158.5.P36 1986 500 86-8165
ISBN 0-13-835380-8

Manufactured in the United States of America

10 9 8 7 6 5 4 3 2 1

First Edition

In loving memory of
Maria, my mother

Acknowledgments

I am grateful to Mary Kennan, my editor at Prentice Hall Press, for inviting me to write this book and for her helpful guidance throughout its completion. Many thanks go to my dear friend George Fichter, who, besides checking and editing the manuscript, was a constant inspiration and source of encouragement. Thanks also to Cissie Ogle, who typed the manuscript, and to Stetson University for helping me complete the manuscript with a summer grant. Their contributions helped to make this voyage a splendid one for me and I hope equally enjoyable for you.

Contents

Preface

In his classic essay "The Two Cultures" (1959) C. P. Snow warned about the growing gap between those who know science and those who do not. He predicted that if this gap were to go unchecked it would lead to an intellectual disaster. A few years earlier Albert Einstein had pointed to the same problem, urging a democratization of science. He wrote of the importance that the general public be given an opportunity to experience consciously and intelligently the efforts and results of scientific research. Einstein believed that the restriction of this body of knowledge to a small group of experts can deaden the philosophical spirit of society and lead to spiritual poverty.

More and more people today find science mysterious, incomprehensible, and even threatening—precisely what Einstein and Snow warned us about.

The purpose of this book is to take a discovery tour of some of the latest topics in science and technology. It's a voyage with interesting stops along the way. The reader is not required to have a scientific background to read and understand this book. The topics, presented in nontechnical language, are:

1. The Nature of Matter
2. Laser—The Miracle Light
3. Nuclear Energy
4. The Silicon Chip
5. Medical Technology and Genetic Engineering
6. Relativity
7. The Cosmos
8. Space Colonization

Besides the description of the topic, each chapter contains relevant historical and socioethical elements, which I strongly believe are indispensable parts of the teaching of science.

The historical element reminds us that our interpretation of nature was, is, and very likely always will be in constant flux, for scientific theories have always been updated or even replaced throughout history. Such an awareness brings out a sense of humility and constructive skepticism that can safeguard us from the evils of arrogance and scientific dogmatism. More, learning about the people who gave us science—the human effort, pain, and jubilation that surround the development of every scientific theory—adds a touch of warmth and humanity to an otherwise dry and impersonal discipline.

The socioethical element reminds us that science has an impact on society and on our lifestyle. Science is advancing at such a fast rate that our greatest challenge is to learn how to adapt quickly and to predict and control these changes actively as much as possible.

The objectives and style of *The Splendid Voyage* are much in tune with the words of Nobel laureate physicist I. I. Rabi:

> Science is an adventure of the whole human race to learn to live in and perhaps love the universe in which they are. To be part of it is to understand, to understand oneself, to begin to feel that there is a capacity within man far beyond what he felt he had, of an infinite extension of human possibilities.... I propose that science be taught at whatever level, from the lowest to the highest, in a very humanistic way. It should be taught with a certain historical perspective, with a certain philosophical understanding, with a social understanding and a human understanding in the sense of biography, the nature of the people who made this construction, the triumphs, the trials, the tribulations.[1]

This book originated from a one-week course titled "The Frontiers of Science," which I offered through the Division of Continuing

Education at Stetson University in Florida, in the summer of 1983. This was part of the Elderhostel Program, which caters nationwide to retired citizens sixty-five years old and older. My greatest challenge in this course was how to explain topics such as relativity, lasers, and nuclear energy to people who had little or no science background.

THE SPLENDID VOYAGE

Chapter One
The Nature of Matter

What is common to a star, a rose, an ice cream cone, and a computer? At first glance, nothing. But all matter, despite its rich and colorful variety, is made of only a few common particles. Up to a few years ago these were known to be the electrons, protons, and neutrons; today we know that protons and neutrons are made of smaller units of matter—quarks.

The atomic theory of matter, first proposed by Democritus in the fifth century B.C., went through periods of popularity and rejection and was often associated with religion and art. Tapping the unveiled secrets of matter produced monstrous spin-offs, like the atomic bomb, as well as invaluable benefits, like television, nuclear medicine, the transistor, and the laser.

Seeking the secrets of the nature of matter is an exciting adventure of the human intellect that is at least twenty-five hundred years old. Philosophy was the first method used, followed by alchemy, a mystical, semireligious pseudoscience. Today we use the formidable alliance of theoretical physics and atom smashers to discover the most intimate secrets of matter. At first this was done by only a few people, with equipment worth only a few hundred dollars. Similar experiments today demand a large community of scientists and engineers operating with million-dollar budgets.

The discoveries of the last sixty years have revealed the fascinating world of the atom. The secrets of this microcosmos were impossible to discover using traditional, Newtonian physics. A bold new physics was invented, based on concepts such as quantum mechanics, wave-particle duality, and the uncertainty principle. The atom was no longer a minuscule solar system, but a fuzzy assembly of "clouds of probability." Questions such as "Where is the electron?" were no longer valid. The nucleus was no longer made of protons and neutrons alone but was cemented with a powerful nuclear force playing a ball game of pions. The hundred or more other exotic particles discovered by smashing the nucleus caused confusion at first, but later on the quark model brought order once again—at least for the present.

The quest continues without the certainty of ever reaching a final answer. What is the purpose of it all? Are the answers worth the billions of dollars spent each year? Do we really have to know the answers? Maybe not.

But we must definitely continue to ask these questions. Curiosity has been our greatest aid in reaching a higher plane of existence. Without it, we cease to advance. But, more than that, this question is special to all of us. We are all made of matter, and so is our world. Knowing the secrets of our material makeup is really knowing a little more about ourselves. Maybe this is why early Greek philosophers asked the same question, with the same passionate curiosity.

The Atomic Theory

Crude explanations for many of nature's phenomena and mysteries were offered even by our most primitive ancestors, but true science begins when the answers are not in terms of gods and demons but rather in terms of laws based on rational reasoning. This may have come about when observation enabled people to predict with accuracy and confidence many natural phenomena. Historians trace these beginnings of science in ancient Greece to around the sixth century B.C.

A group of philosophers who lived before Socrates, therefore known as pre-Socratics, sought scientific answers to many fundamental questions—such as the origins of humans, the movements of the stars, and the size and shape of the earth. The most intriguing question, which captured the interest and imagination of generations of Greek philosophers, was on the composition of matter. What is the prime element, if any, that makes up the materials of the world? What are the laws that govern the making, transformation, and decomposition of materials?

The first of the pre-Socratic philosophers was Thales of Miletus (625–545 B.C.), acknowledged as the first human being who can be rightly called a man of science. A merchant who traveled extensively, he learned geometry in Egypt and studied astronomy in Mesopotamia. He proposed that the prime element of all the material world was water. He must have learned of its vital role in the Egyptian and Babylonian stories of creation, but other, more convincing evidence must have captured his imagination. Water is omnipresent—in the sky, earth, and oceans. It is a major fluid in plants and animals, and without it life is impossible. It is in a constant, life-giving cycle, and was then the only material known to exist in all three states: as solid, liquid, and gas.

Anaximander (611–547 B.C.), the pre-Socratic philosopher who developed a theory of evolution, proposed that the world was built from a universal medium entirely without definition or character. Anaximenes (550–475 B.C.) chose air as his prime element. In successive condensations, according to Anaximenes, air could become wind, cloud, water, mud, earth, and stone; in successive degrees of rarefaction it could become ether, the incorruptible substance thought to make up all celestial bodies, and finally pure fire.

Heraclitus (550–475 B.C.) chose fire as the prime element, because it fit well with another aspect of his philosophy, on the controversial question of changes in nature. Some believed that, despite minor changes, the overall material world remained unchanged. Parmenides (546–506 B.C.), who supported this theory, believed that all there is had existed and would exist for eternity. He went on to propose that matter could neither be created from nothing nor totally be destroyed—stating two thousand years ago the modern physics concept of the law of conservation of matter. The principle of no change is the essence of a modern cosmological theory—the steady-state theory, which lately has been overshadowed by the popular big-bang theory.

Heraclitus led those who opposed the views of Parmenides, believing instead in a nature that was in constant flux. For Heraclitus fire, with its ever-changing flame and its power to transform materials, was both the symbol and the evidence for his theory of eternal flux.

Around the fifth century B.C. two important but opposing theories of matter were introduced. Empedocles (484–424 B.C.) suggested that the material world was made of not one but four basic elements: air, water, fire, and earth. Each was associated with two of four basic qualities: cold, wet, hot, and dry. Air was wet and hot, water was wet and cold, fire was dry and hot, and earth was dry and cold. The following diagram illustrates the relative position of qualities and elements.

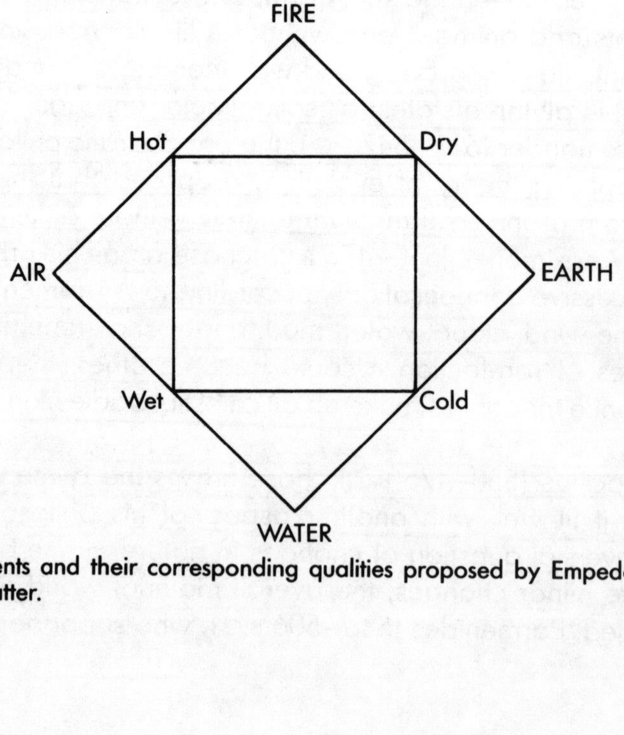

The four elements and their corresponding qualities proposed by Empedocles to make up all matter.

These four elements, combined in an infinite number of variable proportions, could form all matter, animate and inanimate. Empedocles explained the coming together or breaking up of these elements as the result of a continuous battle between nature's two major forces of "love" and "strife," representing attraction and repulsion.

Democritus (460–370 B.C.) developed the other major theory. Expanding on the ideas of his teacher, Leucippus, he proposed that all matter was made of minute indivisible particles, which he called atoms (from the Greek word *atomos*, meaning indivisible). Although little of his writings survived, his theory was described well by other philosophers, particularly Epicurus (341–270 B.C.) and the Latin poet Lucretius (100–55 B.C.). Poetry is the last form of writing in which one would expect to read about an atomic theory, but in his masterpiece, *De Rerum Natura (On the Nature of Things*), Lucretius describes the atoms of Democritus not only accurately but beautifully. He depicts the primordial particles arranging themselves to form all types of matter. Through vast periods of time and after trying every kind of union and motion, these primordial atoms were "impelled by blows and moved by their own weight... flocking together, oftentimes to become the rudiments of mighty things of earth, sea and sky, and the race of living creatures."[2] According to Lucretius, the concentration of these particles and how much space was between them determined whether the material formed was hard, such as rock or metal, or light, such as water or air.

Despite its primitiveness, the atomic theory of Democritus bears striking similarities to modern physics concepts. For example, matter was thought to be eternal; all atoms were made of a common substance and were in ceaseless motion. In other ways the theory was crude and simplistic but imaginative. It was suggested that the atoms of different materials were of different shapes and sizes and that their shape was responsible for their physical and chemical properties. The atoms of most liquids were visualized as round and smooth, so that they could glide over one another easily, whereas in solids the atoms were spiky and rough, clinging to one another with hooklike tentacles to ensure strength and hardness.

Taste also was explained in terms of atom texture. A sharp and sour taste, like that of vinegar, was caused by sharp points that prickled the tongue; sweetness was due to atoms with very smooth surfaces.

Democritus recognized the critical need for vacuum, or "void," in his theory. Like pebbles in a jar, atoms could not fill all the available space no matter how closely they were packed. The idea of a permanent void was anathema to philosophers like Aristotle (384–322 B.C.), who considered it totally unnatural. But the atomic theory was opposed by other, stronger arguments. Democritus proposed that even the mind and the soul were made of atoms. He described a world that was purely materialistic, formed by accident and having no particular purpose. These were concepts to which most Greeks could not subscribe. They refused to believe that their most precious and mysterious faculties—mind and soul—could be reduced to a random rearrangement of atoms.

Aristotle, whose theories dominated Western thought for almost two thousand years, attacked the atomic theory in many of his scientific works. Influenced in his antimaterialism by his great teacher, Plato, he accepted instead the four-elements theory of Empedocles.

Although the atomic theory was not entirely abandoned, up to A.D. 1600 it had only a few supporters, not just because Aristotle rejected it but also because to believe in a world without spirit, order, or purpose was blasphemy to Jews, Christians, and Muslims. The erratic motion of atoms was not a satisfactory replacement for the grand design of a creator. Theologians of these religions considered those who subscribed to the atomic theory materialists and atheists.

The glorious years of Greek ideas, theories, and debates on the nature of matter were followed by centuries of relative stagnation. The only activity that continued to be concerned with the secrets of nature and the transformation of matter was alchemy, a pseudoscience that first appeared in the second century A.D. Alchemy was much more than a primitive form of chemistry. Often secretive, it combined the craft of metallurgy with Aristotle's ideas about matter and with elements of philosophy, astrology, religion, and mysticism.

Alchemists believed in the unity of matter, that metals were related to the planets, that they possessed a spiritlike nature, and that they were going through an evolutionary process from their most primitive state to their most refined—gold. Alchemists believed that since nature always strives toward perfection (an Aristotelian concept), there must exist an agent that can change, or transmute, ordinary metals into precious metals. Alchemists searched for hundreds of years for that agent, known as the "philosopher's stone," that could cure the imagined diseases of metals and elevate them to silver and gold, their most healthy and pure states.

They believed that such an agent would also cure human diseases and thus restore and preserve health and even youth. Viewed in this manner, the agent was known as the "elixir of life." Some alchemists believed it was cinnabar, the powdery red mercury sulfide that when heated converts into mercury, a silvery liquid metal, and sulfur, a dry yellow powder.

The Arabs, who were introduced to alchemy after capturing Alexandria in A.D. 640, spread it to Western Europe around the twelfth century. They adopted the four-elements theory of matter and proposed that all metals had their origin in mercury and sulfur. The contributions of Arab and European alchemy to science were minimal. Apart from discovering a few new metals and isolating some acids, their major contribution was in the new chemical apparatus and techniques that were developed, such as distillation flasks, calcination, fermentation, and sublimation. In the hands of scientists who were free from magic and superstition, these would become invaluable gifts in unveiling the mysteries of matter.

Pierre Gassendi (1592–1655), a French priest and philosopher, prepared the way for the revival of the atomic theory by removing the stigma of atheism associated with it. He accepted atoms as God's creation. He supported the science of the atomic theory, but rejected the ideas that there is no divine providence and that the soul is not immortal.

By the end of the seventeenth century the popular works of

Copernicus, Galileo, and Newton had undermined the authority of Aristotle to an extent that the atomic theory could be reconsidered. Galileo and Francis Bacon laid the foundations of the modern scientific method, wherein observation, mathematics, and quantitative experiments have replaced the scholastic method of philosophy and the authority of ancient writers.

The magic could glitter no more; alchemy was no match for the power of the new science. The attack was on, and alchemy was dying a slow death. In 1661 the Anglo-Irishman Robert Boyle (1627–91), a founding member of the Royal Society in England, attacked alchemy in his scientific classic *The Sceptical Chymist.* Although skeptical about his own views on matter, Boyle gave a new definition to the term *element.*

> I mean by elements...certain primitive and simple or perfectly unmingled bodies; which not being made of any other bodies, or of one another, are the ingredients of which all those called perfectly mixt bodies are immediately compounded, and into which they are ultimately resolved; now whether there be any one such body to be constantly met with in all, and each, of those that are said to be elemented bodies, is the thing I now question.[3]

When, in 1789, the French chemist Antoine Lavoisier (1743–94) published his book *Elements of Chemistry*, chemistry was finally established as a legitimate modern science. Mysticism was now replaced by a new powerful tool, quantitative analysis. The measurements of chemicals before and after a reaction gave an insight into the properties and nature of matter. Chemists realized that some materials, which they called compounds, could be separated into component materials, but others, called elements, could not. Elements were seen as the fundamental blocks of matter.

The Quaker scientist John Dalton (1766–1844), a Manchester schoolteacher, considered meteorology his first love and kept a daily weather record for forty-six years, yet his most notable contributions were in chemistry, his second love, and atomic theory.

In his book *New System of Chemical Philosophy* (1808) Dalton made the novel suggestion that different chemical elements were made of different atoms, so there were as many different atoms in nature as there were elements. The atoms of different elements differed in size and mass, and during chemical reactions atoms were neither destroyed nor created, only rearranged.

Thanks to Dalton, the atomic theory of matter became a powerful new way to explain many chemical and physical phenomena. But Dalton also showed the way to the future. He attempted to classify atoms by weight, giving hydrogen, the lightest atom, the atomic weight 1. He even proposed a diagrammatic representation for each atom, using circles with lines and dots. (Today elements are represented by either their initial letters [O for oxygen, Th for thorium] or some combination of letters from their Latin names [Cu for copper, from *cuprum*, an ancient name for Cyprus; Au for gold, from *aurum*], a system proposed by the Swedish chemist Jöns Jakob Berzelius [1774–1848].)

While the science of chemistry was enriched even more by the works of such scientists as Amadeo Avogadro (1776–1856), Berzelius, Joseph Louis Gay-Lussac (1778–1850), and Humphry Davy (1778–1829), clouds of confusion appeared in the second part of the nineteenth century. By 1872, sixty-three elements had been identified. In addition to their atomic weight, other constants were measured—density, melting and boiling points, and electrical conductivity. The many chemical and physical similarities suggested that elements could be grouped into families with common properties. But which property should be the yardstick of classification? Many schemes were suggested; the most successful one was that of the Russian chemist Dmitri Ivanovich Mendeleev (1834–1907), proposed in 1869. He arranged the elements in a table in order of increasing atomic weight, in such a way that columns contained elements with the same chemical properties. This arrangement is today known as the periodic table of the elements. Mendeleev suggested that the gaps in the table represented undiscovered elements, and he predicted their properties accurately.

Early Twentieth-Century Breakthroughs

Scientists were filled with a sense of confidence and euphoria in the latter part of the nineteenth century. The power of Newtonian mechanics was awesome. All motion could be described accurately with only a handful of equations. From orbiting planets to falling apples or cannonballs, Newton's laws reigned supreme. The new phenomena of electricity and magnetism were described in a set of equations by the brilliant Scottish physicist James Clerk Maxwell (1831–79). Many saw the universe as a giant clockwork, whose mechanism and operation could be explained easily by the laws of physics. Apart from the need to unify all phenomena under one theory, physicists felt that all that was left for them to do was add another decimal point to their measurements. Many went so far as to advise young scholars to stay away from physics, for there were no remaining secrets to discover in nature. Little did they know.

Modern physics begins in 1896, a vintage year for science. Around that time three great discoveries shook the foundations of physics, particularly those based on the atomic theory. A major scientific controversy that took on almost nationalistic overtones concerned the nature of cathode rays, the colorful tracks of electricity formed in a low-air-pressure glass tube. Some thought they were rays of light, but others, like William Crookes (1832–1919), suggested that they were molecules carrying electric charge. After all, they could be deflected by a magnet.

In 1896 the German physicist Wilhelm Roentgen (1845–1923) discovered that energetic cathode rays striking a metal can produce a new kind of ray, invisible and extremely penetrating. Roentgen named these X-rays, and for his discovery he was awarded the first Nobel Prize in physics, in 1901. Besides contributing greatly to medicine, X-rays triggered a new spirit of inquiry in the stagnant world of science.

Within weeks of the discovery of X-rays, the French physicist

Henri Becquerel (1852–1908) discovered another kind of invisible radiation, emitted spontaneously by a piece of uranium. To Becquerel, this was an outright violation of one of nature's most sacred laws—the law of conservation of energy. Since energy cannot be created out of nothing, where was this continuous stream of radiation coming from? Pierre (1850–1906) and Marie Curie (1867–1934), who worked with Becquerel, isolated two other materials, radium and polonium, that emitted the same type of invisible radiation, called radioactivity. All three shared the Nobel Prize in physics in 1903.

The answer to the question of the source of energy in radioactivity would be provided by Albert Einstein (1879–1955) when, in 1905, he published his famous paper on relativity. In that short paper Einstein introduced a most unconventional concept—that matter and energy are manifestations of the same thing. His famous formula $E = mc^2$ (E stands for energy, m for mass, and c for the speed of light) became the symbol of twentieth-century physics. The atoms of radioactive substances get smaller, spontaneously emitting particles and energy, or radioactivity. According to the formula, radioactivity comes from the conversion of some mass into energy.

In the meantime, at the Cavendish Laboratory of Cambridge University in England, J. J. Thomson (1856–1940) was on the verge of a major discovery. In 1896, after years of work on the electrical conductivity of gases, he carried out a classic experiment that unveiled the mystery of cathode rays. He concluded that these rays were negatively charged particles, and from measurements of their deflection in electric and magnetic fields he calculated their speeds and their ratio of mass to charge (m/e). The results were amazing. These particles had speeds of thousands of miles per second, and, most disturbing, their mass was 1,800th that of the smallest atom, hydrogen. This suggested that the atom of Dalton was divisible into smaller parts, which Thomson was reluctant to accept. But after many more experiments, the extreme consistency of the results left him with no choice but to admit to the divisibility of the atom.

"There is no escape from the conclusion that we are dealing with particles that are far, far lighter than atoms,"[4] he reported.

Thomson had discovered the electron, the smallest of the three prime atomic particles, and for this he won a Nobel Prize in physics in 1906.

In 1904 Thomson suggested a model for the atom in which both the mass and the positive electric charge were represented by a sphere containing an equal amount of negative charge in the form of electrons. This was called the "plum pudding" model, because it resembled a pudding with the electrons scattered in it like raisins.

Only five years later, in 1909, the plum pudding model failed a crucial test carried out by Thomson's most brilliant student, Ernest Rutherford (1871–1937). Rutherford, with two of his students, Johannes Geiger and E. Marsden, fired alpha particles at thin gold foil to study their deflection. Alpha particles are emitted by many radioactive elements, and they carry a double positive charge. (They were identified later as helium nuclei.) Since the Thomson model assumed that the positive charge was distributed throughout the body of an atom, any deflections of alpha particles by electric repulsion were expected to be small.

Rutherford and his team were stunned by two unpredictable results. More alpha particles than predicted could shoot right through the gold foil, as if indeed the atoms were made of empty space. Those that did not were deflected to a variable degree, as measured with a fluorescent screen. Some were even deflected right back to where they originated. For alpha particles traveling at about ten thousand miles per second to suffer such a deflection, they must encounter a powerful force of repulsion. Such a force could only be electrical in nature and could be generated only if all the mass and positive charge of the atom were concentrated in a tiny central core—the nucleus. Recollecting those exciting moments, Rutherford wrote years later:

> It was quite the most incredible event that has ever happened to me in my life. It was almost as incredible as if you fired a 15-inch shell at a piece of tissue paper and it came back and hit you. On consideration, I realized that this scattering backward must be the result of a single collision, and when I made calculations I saw that it was impossible to get anything of that order of magnitude unless

> you took a system in which the greater part of the mass of the atom was concentrated in a minute nucleus. It was then that I had the idea of an atom with a minute massive centre, carrying a charge.[5]

In 1911 Rutherford published a new model of the atom. Its central core, the nucleus, occupied only a minute portion of the volume, yet it held 99.9 percent of the atomic mass. Both these ideas hold true today. The electrons were thought to circle the nucleus in one shell, an idea that was soon modified. To get a real appreciation of atomic dimensions, visualize a simple atom, like hydrogen, enlarged to the size of a football field. The nucleus, at the center of the field, is the size of a letter "o" in this text, and the electron orbiting at the perimeter of the field is the size of a minuscule dot. The rest of the atomic volume, the football field, is empty space.

The discovery of the nucleus was a crucial step in the development of the atomic theory. The nucleus was recognized as the source of radioactivity and later on of nuclear energy. It is somewhat ironic that the formula that explains both, $E = mc^2$, had been proposed by Einstein six years before the nucleus was discovered.

But even Rutherford's model had a basic flaw. According to Maxwell's electromagnetic theory, an orbiting electron should emit electromagnetic waves, like a tiny radio transmitter. This continuous loss of energy should cause its orbit to shrink so fast that in less than a second the electron would spiral into the nucleus. This problem of explaining atomic stability was a clear indication that the old physics of Newtonian mechanics had its limitations. There was a need for new ideas and a different approach to atomic physics. These changes could come only from young (and brave) physicists whose minds had had no time to harden in the tradition of the old physics. The Danish physicist Niels Bohr (1885–1962) was one of these young lions.

After only a year as a visiting physicist at Rutherford's laboratory, in 1913 Bohr proposed a stable atomic model. In this model, the electrons go around the nucleus in well-defined circular orbits, like planets orbiting the sun in a miniature solar system. When atoms absorb energy, some of their electrons jump to higher orbits (excited states). In less than a billionth of a second, they fall back to their original orbit (ground state) in one or a series of downward steps through other excited states. In each "down" step a packet of light energy, or photon, is released. The "height" of each step is a measure of the energy of the emitted photon, which in turn corresponds to its color—blue for a tall step and red for a short one. Each element emits a unique combination of colors, its characteristic spectrum.

In the Bohr model, only discrete amounts of energy can be absorbed or emitted by the atom. When in a stable state, the atom cannot emit energy.

The success of this model is in its ability to explain the spectra of elements such as hydrogen, helium, and mercury. Although obsolete by today's standards, the Bohr model is an important part of the atomic folklore. It has a unique place in history because it contains novel ideas of the young and still untested new physics—quantum mechanics.

In 1900 Max Planck (1858–1947) proposed that radiant energy can be emitted and absorbed by atoms only in discrete amounts, called quanta (*quantum* is Latin for "how much"). Einstein used this idea to explain the photoelectric effect, the production of electricity by light shining on a metal. Einstein considered light to be made of packets of small energy—the quanta of light, photons. For this work Einstein won the Nobel Prize in physics in 1921. This and Bohr's explanation of spectra were indisputable proofs of the power of the new physics.

The concept of the quantum led to other new ideas that redefined basic concepts of physical reality. In 1924 the French physicist Louis de Broglie (1892–) proposed in his doctoral thesis that moving particles have wave characteristics. Bizarre as it may sound, this can be proven experimentally. When fast electrons are shot through narrow slits, they can produce interference, an effect associated only with waves.

De Broglie's theory suggests that moving electrons, including those that orbit the nucleus, can be treated both as particles and as waves, a concept that is now known as the "dual nature of matter." Erwin Schrödinger (1887–1961) derived a mathematical model to account for that behavior. The famous equation that bears his name defines the wave properties of electrons and also predicts their particle behavior:

$$-\frac{h^2}{2m}\frac{d^2\psi}{dx^2} + V\psi = ih\frac{d\psi}{dt}$$

For a moving particle of mass (m), its kinetic energy $\left(-\frac{h^2}{2m}\frac{d^2\psi}{dx^2}\right)$ plus its potential energy ($V\psi$) is equal to its total energy $\left(ih\frac{d\psi}{dt}\right)$, where

$h = \frac{h'}{2\pi}$ (h' = Planck's constant; $\pi = 3.14$)

ψ = the wave function of the moving particle (a mathematical expression that represents its wave characteristics)

$i = \sqrt{-1}$

$\frac{d^2}{dx^2}, \frac{d}{dt}$ = mathematical operators.

In 1927 Karl Werner Heisenberg (1901–76) developed the uncertainty principle, for which he won the Nobel Prize in physics in 1932. This states that it is physically impossible to measure precisely the position and the velocity of an object simultaneously. Errors will always accompany any such measurements. As a crude illustration of this principle, consider that we need to know the position of a stationary electron. The smallest "informant" that we can send to find out the location of the electron is a photon. Its collision with the electron is vital to the measurement, yet distorts both the position and the velocity of the electron. This is analogous to a fast-moving marble hitting a stationary billiard ball.

So precision questions—like "Where is the electron?"—are no longer valid in atomic physics. The uncertainty principle forces us to ask instead, "What is the probability of finding the electron in such a volume of atomic space?" Max Born (1882–1970) was the first to introduce the concept of probability into quantum theory. Orbiting electrons are no longer visualized as in well-defined orbits but in "clouds" of probability. The solutions of Schrödinger's equation describe the shapes of these clouds of probability for electrons at different energies. Although these mathematical images of atoms are for consumption only by physicists, they prove that even in its deepest secrets nature is full of symmetry and beauty.

Despite the success and popularity of quantum theory, some physicists rejected parts of it, like the "fuzzy" picture of the atom. Einstein, whose early work strengthened quantum theory, was one

Clouds of probability showing the location of the electron in a hydrogen atom for different energies. These clouds are worked out mathematically, using the quantum theory.

of the skeptics. He could not accept that atomic events were obeying only laws of random behavior. He believed in an orderly universe whose behavior could be described accurately according to natural laws. The exchange of views with his good friend Niels Bohr was intimate and intense. To Einstein's frequent remark "God does not play dice with the Universe," Bohr replied one day, "Albert, stop telling God what to do."[6]

Quantum mechanics has been a successful new way to understand small-scale phenomena in both chemistry and physics. Its power is well illustrated in the words of physicist Victor Weisskopf, who speaks of his "most impressive experience" during World War II:

> We knew that an element which we called plutonium ought to exist. Nobody had seen it.... All we knew was that the charge of its nucleus was 94... and that it had 94 electrons around it. So we sat down and calculated its properties. The fact that this calculation could be done at all shows the tremendous power of quantum mechanics. We could predict that it must be a metal; that it must be brown colored; that it had a particular spectrum; that it had a certain hardness and a certain elasticity. And when we had the first cubic inch of plutonium before us, it turned out that all of this was true.[7]

Atom Smashers, Quarks, and Gluons

The picture of the atom was thought to have been completed when Chadwick discovered the neutron in 1932. The atom was seen as a minuscule solar system, with the nucleus, made of protons and neutrons, at its center and electrons circling in different orbits. Each particle plays a vital part in the stability and the properties of the atom. The neutron has no electric charge, and it is slightly heavier than the proton. Together they generate the "cementing" nuclear force that keeps the nucleus together. Without this force the nucleus would explode instantly, due to the large electrical force with which the positively charged protons repel one another. On the other hand, the electrical force between the positive nucleus and the negatively charged electrons is attractive and keeps the electrons from flying away. The electrons, which are 1/1,840 the size of the protons, are arranged in orbits, each of which can accommodate only a certain number. The physical and chemical properties of the atom are determined by the number of electrons, or indeed the "empty spaces," in the outer orbit, a concept known as valence. Quantum mechanics describes these orbits mathematically as clouds of probability.

Atoms do not stay stable forever. They can all undergo a variety of changes. The ones with heavy and unstable nuclei, seeking a more stable state, "shrink" by emitting energy and particles, or radioactivity. Atoms can lose one or more electrons by collision, remaining temporarily positively charged. In this state they are known as ions.

Atoms can also be forced to change when particles are injected into their nuclei. The first such change was made by Rutherford in 1919, when he sent an alpha particle (made of two protons and two neutrons) into a nitrogen nucleus and changed it into an oxygen atom. The dream of alchemy had at last come true. In principle, any atom, appropriately bombarded, can change into any other atom, including gold, but the cost for even tiny amounts of a new element is astronomical.

The mysteries of the nucleus have captivated physicists since the 1930s. What really goes on in that tiny space packed tightly with matter? How do the nucleons (protons and neutrons) behave? A number of nuclear models have been proposed, each using a different part of physics to explain a different nuclear phenomenon. In the "shell" model, quantum mechanics is used to describe the arrangement of nucleons in shells within the nucleus and to explain why some atoms are stable and others not. The "liquid drop" model, on the other hand, explains nuclear properties in terms of classical physics. The nucleus is viewed as an oscillating drop of liquid with its nucleons in constant random motion, like the molecules in a liquid. This model explains how a nucleus splits during fission and how gamma rays, the most penetrating type of radioactivity, are generated by nuclear oscillations.

But of all the nuclear mysteries, the nuclear force is the most fascinating. It is the strongest force in the universe, yet it can be activated only when nucleons are at distances shorter than 10^{-24} meter. If this force were to be inactive for only a billionth of a second, the whole universe would disintegrate instantly in a titanic explosion, with every nucleus flying apart because of the electric repulsion of its protons. But how does one nucleon know of the existence of another? How, indeed, is this strong force communicated from nucleon to nucleon?

The Japanese physicist Hideki Yakawa (1907–) used quantum mechanics to develop in 1935 a theoretical model in which the nuclear force is communicated from nucleon to nucleon via a particle, the pion (or pi-meson). He predicted its mass to be three hundred times that of the electron. Yakawa's theory is a milestone in the history of nuclear physics. The nucleus was no longer the exclusive habitat of only protons and neutrons. Experimental physicists rushed to develop techniques for "capturing" the pion and any other new nuclear particles. This was the beginning of a new era—the era of large, powerful, and expensive atom smashers.

Smashing the nucleus and examining its fragments required not only vast sums of money but also a large group of scientists. But this new facet of physics research did not take American science by surprise. The experience of the Manhattan Project in making the atomic bomb (1939–45) was a fine example of a large and well-coordinated scientific research effort.

Making sense out of the fragments of a smashed nucleus is analogous to trying to understand the working mechanism of a watch by blasting it with a bullet and examining the fragments while they are still in flight. Crude as it may sound, this technique shed light on many secrets of nuclear structure. The complete task is divided into two stages. In the first stage, the "bullet" is prepared and fired. In the second stage, the nuclear fragments are photographed and then interpreted.

The first bullets used to bombard a nucleus were alpha particles emitted by radioactive materials. Since they have low energy, they can penetrate only small nuclei. When heavier elements are used as targets, the repulsive electrical force between a large nucleus and an approaching alpha particle is strong enough to prevent an encounter. In the early 1930s physicists realized that devices had to be employed to pump up the energy not only of alpha particles but also of all other nuclear bullets—such as electrons, protons, and a few light ions. Since all of them carry an electric charge, the easiest way of speeding them is to accelerate them with attractive electric fields. Since their motion has to be free of any obstruction, including air, their flight paths are inside pipes with an extremely good vacuum.

The first successful atom smashing was carried out in 1932 by Cambridge University physicists John Cockcroft (1879–1967) and E. T. S. Walton (1903–). At about the same time, American physicist Robert Van de Graaff (1901–67) developed a machine (which bears his name) that could generate the millions of volts of electricity needed to pump up the energy of nuclear bullets. A conveyor belt picks up static electricity created by friction and deposits it on a large spherical metal shell. After some time of operation, the shell can be charged up to ten million volts, which is enough to throw a lightning bolt. In 1933 the Van de Graaff generator was used to smash a proton into a hydrogen nucleus, which is made up also of one proton. The result of this proton-proton collision provided the first experimental evidence of the nuclear force. A more exciting accelerator was the cyclotron, devised by Ernest Orlando Lawrence (1901–58) in 1931. This was a circular vacuum chamber that employed a high-frequency magnetic field rather than an electrical field to accelerate charged particles in an expanding spiral.

Accelerators were made in a great variety of shapes and sizes after World War II. Some were doughnut-shaped (bevatron, synchrotron) and others were straight (LINAC, linear accelerator). Both types used high-vacuum pipelines and a combination of electrical and magnetic fields.

Each accelerator had a limit to the maximum energy that it could impart to a nuclear bullet. The unit of energy is the electronvolt (ev), which is the energy acquired by an electron when it is attracted by a voltage of one volt. Early accelerators had a capacity of only a few thousand electron volts (Kev). Later they grew stronger, reaching a million electron volts (Mev) and even a billion (Bev). Some of today's superaccelerators can reach energies on the order of Gev (1 Gev = 1,000 Bev).

These fantastically high energies mean that the bullet particles can smash larger nuclei into even smaller particles. The speeds of these nuclear bullets are astronomical. For example, a proton at 10 Bev travels at 2.998×10^8 meters per second (299,800,000), which is 99.996 the speed of light.

The nuclear fragments can be detected because they ionize the molecules in their path. These ions can be made visible as tiny bubbles in a liquid, as minuscule droplets in cold and moist air, or as a series of small sparks in multiple layers of electric plates. The cloud chamber, built in 1911 by the Scottish physicist C. T. R. Wilson (1869–1959), was the first such detector. Both the instrument and its operation are extremely simple. Retracting a piston makes the moist air or alcohol in a glass container suddenly expand. Since expansion causes cooling, droplets form. They tend

An aerial view of the Fermi accelerator near Chicago, a 0.64-mile-radius giant atom smasher. (Courtesy Fermi National Accelerator Laboratory)

to form around the ions in the tracks of radioactivity emitted by a source, so the footprints of radioactivity appear as white lines that are really made by tiny droplets, like miniature pearls on a string.

The most popular detector is the bubble chamber. Invented by Donald A. Glaser (1926–) in 1952, it is made of a strong steel tank with large thick glass windows. Cooled down to −250°C, it is filled with liquid hydrogen. The conditions are so critical that any small decrease in pressure will cause the formation of bubbles. When nuclear fragments fly through the liquid hydrogen, they leave ions on their tracks. Each ion becomes a tiny bubble when suddenly the pressure is slightly reduced by a piston. Cameras photograph the tracks as they bend and spiral under the influence of a large magnetic field.

To the trained eye of a physicist, these abstract curves and spirals are a treasure trove of information. After measuring the angles of deflection, the spiral radius, and the length of the track,

Bubble chamber tracks of subatomic particles. When the length and curvature of these tracks are subjected to conservation laws of physics they yield a treasure of information about the structure of matter. (Courtesy Fermi National Accelerator Laboratory)

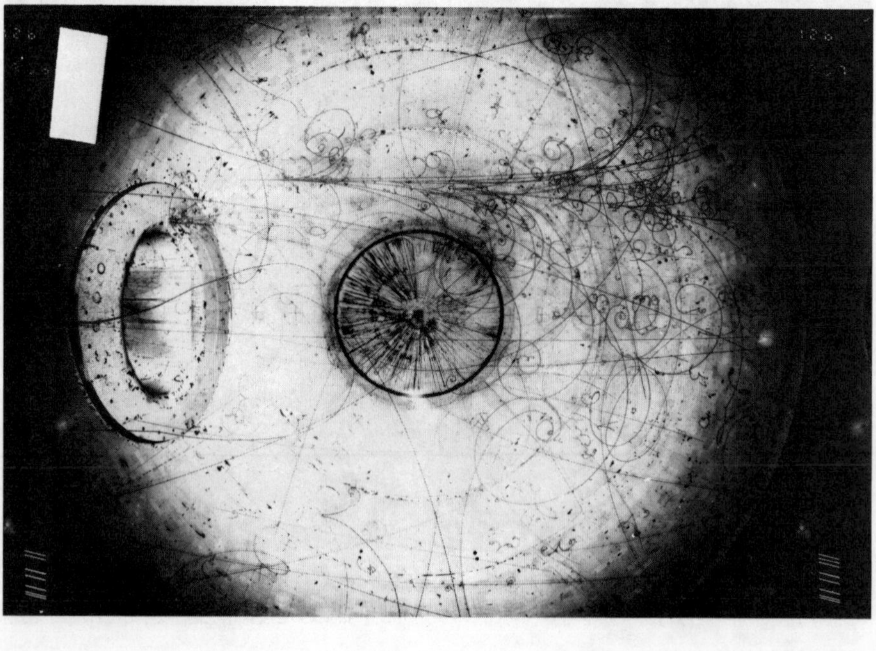

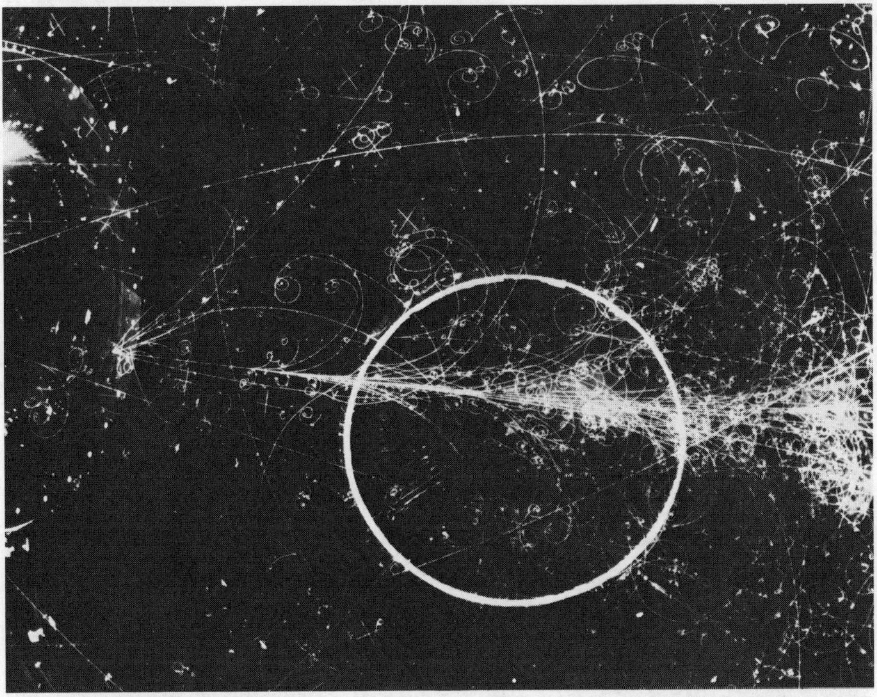

a physicist applies such laws of physics as the conservation of energy, momentum, and electric charge to calculate the mass, charge, and lifetime of subnuclear particles.

Over two hundred new particles were discovered from the millions of bubble chamber photographs taken during the late 1950s and early 1960s. Most do not really exist in ordinary matter but are created out of the energy of collisions, according to Einstein's concept that matter and energy are equivalent. Many of these new particles are so peculiar that it is as if they were from another universe. Among them is antimatter—particles very similar to matter but with opposite electric charge. Antimatter was predicted as early as 1928 by Paul Dirac (1902–84), a young physicist at Cambridge working on a theory that combined relativity with quantum mechanics. He was so shocked by the implications of his mathematics that he thought at first that he had made a monstrous error in his calculations; his conclusion was that for every particle in nature there must exist a corresponding antiparticle. The antiparticle for the electron, the positron, was discovered by Carl Anderson (1905–) in 1932 in photographs of cosmic rays. Antiparticles are not found floating around, because when they meet their corresponding particles they annihilate each other, giving off energy. There are reasons to believe that parts of the universe are made of antimatter.

The list of the "fundamental" particles grew larger each year, causing mixed excitement, confusion, and embarrassment in physicists. The very fabric of the atomic theory was falling into disarray. The question of what makes a fundamental particle was under scrutiny. J. Robert Oppenheimer, commenting on the proliferation of particles, suggested that a Nobel Prize should be awarded to the first physicist who did not discover a new particle in a given year.

Even naming these particles was a problem. The Greek alphabet was used several times over. Enrico Fermi said, "If I could remember the names of all these particles, I would have been a botanist."[8] To add to the difficulty, some of these particles could exist for only incredibly short periods of time. Most of them could live a relatively long life of up to a billionth of a second, but a few others could survive only for 10^{-24} second (1 divided by a trillion trillion). Some particles cannot even be detected, because they are difficult to capture. These are the neutrinos—"ghostly" particles that go through matter as if it was not there, at speeds nearing that of light. They are thought to be everywhere and may be the most abundant particles in the universe. Physicists are still puzzled by their nature, particularly whether they have any mass.

Seeking order to what physicists called the mess of the "particle zoo" was a primary task in the 1960s. By that time all particles were classified according to their mass and nuclear properties and were relegated to groups, like hadrons, leptons, mesons, and baryons. A new Mendeleev was needed to discover patterns in the complex family of subnuclear particles. The new tools were to be found in the principles of mathematical symmetry of a new technique known as group theory.

In 1961, Murray Gell-Mann (1929–) independently developed a method that could classify hadrons, the heavy particles. Gell-Mann proposed that the gaps in this classification represented particles that had not yet been discovered. The discovery in 1964 of one of these particles, the omega minus, was a triumphant verification of the symmetry theory.

But organizing particles into groups of symmetry was one thing; seeking the elementary building blocks of matter was another. In 1964 Gell-Mann and George Zweig (1937–) independently developed the idea of one unit for all particles, the quark. The name was proposed by Gell-Mann, from *Finnegans Wake* by James Joyce. The model of the quark restored some order and brought relief to particle physicists.

Three quarks were originally suggested, but today it is believed that there must be at least six, called up, down, sideways, charm, beauty (or bottom), and truth (or top.). Each carries a positive or negative electric charge with a magnitude a fraction of that of the electron. All particles except leptons (neutrinos, electrons, and muons) are a combination of two or three quarks. (Antimatter, of course, is made of antiquarks.) A proton, for example, is made up of two up quarks and one down quark (uud), with a total electric charge of $+ \frac{2}{3} + \frac{2}{3} - \frac{1}{3} = 1$, which is indeed the charge of the proton. A neutron is made up of one up and two down quarks (udd), with a total electric charge of $+ \frac{2}{3} - \frac{1}{3} - \frac{1}{3} = 0$.

To explain why some combinations are not allowed, physicists had to endow each quark with another property; it is known as color but has nothing to do with real color. Each quark can come in three different colors: "blue," "red," and "green." The theory that explains how quarks of different colors come together to form particles is known as quantum chromodynamics (*chromos* is the Greek word for color). Only quarks with colors that can add up to "white" form any particle. This is possible in two ways—when all three colors are present or when a quark of one color combines with an antiquark of the appropriate anticolor. For example, for a uud proton, each of the quarks must have a different color.

Although individual quarks have never been found, there is substantial experimental evidence that they exist. At the linear accelerator at Stanford University, physicists fired electrons at 20 Bev energies into protons. Their data suggested that the protons were made of moving parts, which were probably quarks. In an experiment at CERN (the high-energy synchrotron of the European Organization for Nuclear Research in Geneva), physicists studied millions of neutrino-proton collisions and found three proton components, as predicted by the quark model.

But the experimental evidence was not always in harmony with the quantum model. Disturbing results forced theoretical particle physicists back to the drawing board. This is how the last three quarks—charm, beauty, and truth—were added to the list. As for what holds quarks together, theoreticians proposed a "glue" that communicates its binding strength from quark to quark via a messenger, a gluon. Experimental evidence for the existence of gluons was reported in 1979 by physicists at Hamburg. The worrying part is that there must be at least eight different kinds of gluons.

So far the quark model has restored order to the chaotic microworld of particles. It is also very satisfying to know that this universe is literally made of a lot of charm, beauty, and truth. Yet these new quarks, together with the fact that eight gluons are needed to bind them, have rekindled old fears of particle proliferation. How many more quarks and gluons will we discover? Some physicists who are well aware of the sins of determinism are reluctant to accept that we can ever have the final answer. Most enjoy the challenge and hope that it will always be there.

The history of the atomic theory stands out as a human adventure full of excitement and controversy, its impact extending beyond science. It has influenced philosophy, religion, and politics and made possible the nuclear arsenal. Even art has felt its touch. Art historians acknowledge the effect of twentieth-century physics concepts such as quantum theory and the uncertainty principle on the style of some painters. For some, reality was no longer described in clear lines but rather in fuzzy abstract images.

The mysteries and the beauty of the atom, with its own standards of reality, space, and time, were expressed by science and art in a great variety of ways. But the first four lines of William Blake's poem "Auguries of Innocence," echoing the sentiments of Lucretius, sum it all up:

> To see a World in a Grain of Sand
> And a Heaven in a Wild Flower
> Hold Infinity in the palm of your hand
> And Eternity in an hour.[9]

Chapter Two

Laser—The Miracle Light

It has been called a miracle light, the supertool of the twentieth century, light fantastic, and the solution to problems still to come. The laser is all of these and much more. From delicate eye surgery to nuclear fusion, from cutting steel plates to restoring masterpieces of art, the laser is one of the most versatile tools ever invented.

It is used in supermarkets and hospitals, telecommunications, industry, weather forecasting, and the military arsenal, to mention but a few applications. Its amazing capabilities—power and precise direction—derive from the purity of its light. First made in 1960 out of a small ruby crystal, lasers have grown more elaborate, made with many different materials, in different sizes, and with different energy outputs.

There is really nothing magic about the production of laser light. It can even be made naturally, as it is in the atmosphere of Mars. To produce it in the laboratory requires the combination of atomic physics with optics and electronics. This multidisciplinary alliance has become a science in its own right, called photonics.

Early Theories on the Nature of Light

Light has bathed earth throughout its 4.6 billion years of history. For the last 3 billion years it has sustained photosynthesis, a chemical reaction essential to life on earth. The sun was worshipped as a god by many ancient cultures. Some even sacrificed thousands of people to it. Religious and artistic symbolism have related light to God, goodness, and life itself, and associated darkness with evil, death, and doom. In Genesis, God makes light on the first day of creation, after the heavens and the earth.

Like air and water, light is an abundant component of our daily life. We built instruments to measure its speed and study its properties. With the latest device, the laser, we produced light in its purest form. Yet light still holds many secrets.

As in times past, light continues to fascinate, to inspire, and to puzzle. It is the Mercury, god of communications, that has built most of our experiences. It delivers to our eyes messages from the world around us. More than 95 percent of all that we know was brought to us by a ray of light. Artists have often used its optical behavior as intricate parts of their style. Scientists are still puzzled by its nature. Some experiments show it to behave like a wave, and others like a particle. What is light?

The early Greeks believed that it traveled at high speed and in straight lines. They were also aware of refraction, its property of bending when going from one medium to another. Some Greek philosophers explained vision as caused by tiny particles in light.

During the Middle Ages, while Europe stagnated intellectually, the Arab world became the steward of Greek knowledge and was alive with curiosity and achievement. Learning was promoted by the religion of Islam. The period of A.D. 900–1100 was for the Arabs a golden age of literature, mathematics, medicine, and technology. Their most famous scholar in optics was Alhazen (905–1038). In his book *The Treasury of Optics*, he described a variety of optical phenomena, such as the rainbow, halo, and reflection, and, for the first time, the theory behind magnification by a lens. To Alhazen light was a kind of fire, reflected at the spherical edge of the atmosphere.

The years between A.D. 1000 and 1200 are the years of translations, during which Europe laid the foundations of an intellectual revival, later to blossom into the Renaissance. During this period hundreds of Greek and Arabic texts were translated into Latin, the scholarly language at the time. Alhazen's book on optics generated a lot of interest and activity in the study of light. In an intellectual climate where theology reigned as the queen of disciplines, the study of optics was in tune with religion. After all, God was synonymous with light. Since light was also one of God's first creations, it followed that it should be one of the first natural phenomena to be studied.

The study of optics led to the development of some extremely important optical devices. The understanding of the anatomy of the human eye and the physics of lenses was combined with the glass technology imported from Byzantium to produce spectacles. Developed around the thirteenth century, these technological miracles, which to some were the work of the devil, added years of useful life for thousands incapacitated by poor eyesight. The arrangement of lenses in a tube resulted in the telescope and the microscope in the seventeenth century. These instruments unveiled two unknown worlds—the very large and far away, and the very small and near.

As to why it took more than three hundred years to make a telescope or a microscope, the answer is one of scientific methodology. Up to the time of Galileo the study of nature was done either from old texts or by the use of philosophy and theology. Scholars followed the doctrine of Plato and other classical philosophers that the senses were misleading and the mind was mightier.

Explaining the colors of sunlight has been another major avenue of investigation. One of nature's most beautiful sights, the rainbow has fascinated observers since antiquity. Aristotle wrote extensively about it in his book *Meteorologica*. He was correct about the role of moisture, but he explained the production of colors in terms of reflection rather than refraction. To Aristotle the colors were due to the density of the reflecting material.

The French philosopher René Descartes (1596–1650) developed two separate theories of light. In the first, he proposed that light was made of tiny particles in rapid linear motion and that all light's properties could be explained in mechanical terms. In the second, he explained light as a kind of pressure transmitted through a dense yet invisible medium, called ether. Colors were produced by the rotation of ether; red was produced by the fastest rotation and blue by the slowest.

The rainbow colors caused by the transmission of sunlight through a glass prism were explained by many as caused by the thickness of the glass, which was thought to "darken up" the light by different amounts. Slight thickness darkened the light only a little and it became red; a lot of glass and it became blue.

The correct explanation was first proposed by Isaac Newton (1642–1727). He proved that light was actually a mixture of colors; all the glass prism does is separate them. Newton was prompted to study the spectrum of sunlight in his effort to correct the coloring effect of telescope lenses. Although his work was done in 1666, the results were not published until 1704, in his classic book *Optics*.

As to the nature of light, Newton proposed a "corpuscular" theory, in which light rays were made of small particles emitted by shining surfaces. His contemporaries Robert Hooke (1635–1703) and Christiaan Huygens (1629–95) proposed instead a wave theory of light, but Newton's overshadowing influence helped the corpuscular theory prevail. Its first major challenge came from a brilliant English scientist, Thomas Young (1773–1829), who supported the wave theory with convincing experiments.

In 1800 the British astronomer William Herschel (1738–1822) discovered a new kind of light that was invisible to the human eye. He passed sunlight through a glass prism and measured the temperature of the colors of the spectrum. He found that the red was the hottest color. But, stranger than that, his thermometer showed an even greater temperature increase when it was placed outside the spectrum, just beyond the red. These are, of course, the invisible heat rays known as infrared. Later, a similar discovery was made with invisible light beyond the violet, called ultraviolet.

In the meantime, Michael Faraday (1791–1867), whose brilliant experimental work gave us the electric motor and dynamo, noticed a peculiar relationship between light and magnetism. He showed in an experiment that a property of light known as polarization was affected by a magnet, as if light did indeed have magnetic qualities. Maxwell, standing next to Newton and Einstein as one of the most powerful theoretical minds in the history of physics, deduced mathematically that light is a stream of rapid electric and magnetic oscillations. He even predicted that high-voltage electric sparks could generate invisible waves traveling at the speed of light. In 1887 Heinrich Rudolf Hertz (1857–94) verified Maxwell's prediction by creating and detecting radio waves.

Maxwell's theory and Hertz's experiments were powerful arguments for the wave nature of light. The early part of the twentieth century, however, produced its share of proofs in support of the particle nature of light. In 1900 Max Planck attempted to explain the mechanism of heat and light emission by proposing that energy is emitted and absorbed only in discrete amounts, called quanta. Einstein used this concept to describe in 1905 the theory of the photoelectric effect, the production of electricity by a metal bathed in light. In 1923 Arthur Compton (1892–1962) proved experimentally that X-ray photons colliding with electrons bounce off in the same way as billiard balls do, conserving momentum (mass × velocity).

These three great physicists won Nobel Prizes for their work, which can be seen as support for the particle nature of light. Yet the evidence of the wave nature of light is still as strong. So what is light? Can it be a wave one moment and a particle the next? Can it be a combination of both? Or is it something entirely different, still unknown and awaiting discovery? Can we ever know the answer? Physicists everywhere admit in all humility that despite their spectacular achievements light continues to be a mystery.

How a Laser Works

To appreciate the uniqueness of laser light one needs a basic knowledge of the production of ordinary light. All light is produced by energized atoms. When an atom absorbs energy, one (or more) of its electrons is raised from its lowest energy state, known as a ground state, to any of one of a number of excited states. In less than a millionth of a second the electron falls back to its ground state, stopping momentarily at some of the intermediate energy states. At each step down the atom emits a packet of light energy, a photon. The energy of the photon is expressed by the formula $E = hf$, where h is a constant known as Planck's constant and f is the frequency of light, which relates to its color. For example, blue light has a higher frequency than red light. Therefore, a blue photon has more energy than a red one. This, of course, means that in the production of blue light the excited electron falls from a higher-energy step than the one involved in the production of red light.

The de-excitation steps and therefore the colors emitted by atoms are different from one element to another. Thus, the light given off by a material is a combination of colors unique to it. No two materials can give off the same combination of colors. The light from a source can be separated into its component colors by an instrument known as spectrometer; each color frequency appears as a line on a calibrated scale or film. This family of colors, each represented by a line, is known as the spectrum of the element. Because each element has a unique spectrum, scientists use spectra as atomic "fingerprints," identifying atoms by their emitted light. This is the method used to identify the composition of stars.

Visible light is only a tiny fraction of a much larger family of waves known as the electromagnetic spectrum, which contains such waves as radio waves, microwaves, and X-rays, all invisible to the human eye. They all travel at the speed of light, but they differ in frequency and wavelength. (Frequency is the number of vibrations per second, and wavelength is the distance between two adjacent peaks in the wave.) The relationship between velocity, frequency, and wavelength is: wave velocity = wavelength × frequency. Since all electromagnetic waves have the same velocity, it follows that if one wave has a short wavelength, like X-rays, then it has a high frequency and therefore high energy ($E = hf$). If, on the other hand, it has a long wavelength, like radio waves, then its frequency is small.

The following diagram shows the relative position of the major parts of the electromagnetic spectrum. Not all are produced by the de-excitation of electrons in energized atoms. Gamma rays, for example, originate in the nucleus, whereas some infrared waves are produced by atomic and molecular oscillations. Radio waves are produced by oscillations of free electrons.

Gamma rays	X-rays	Ultraviolet	Visible	Infrared	Microwaves	Radar Radio

Short wavelength
High frequency

Long wavelength
Low frequency

The electromagnetic spectrum. Despite their different properties, all these waves travel at the speed of light and are composed of electric and magnetic disturbances. (This is not drawn to scale. For example, the "visible" section of the spectrum is much narrower than it appears here. The boundaries are not as rigidly defined as they are in this diagram.)

Any ordinary light, whether produced by electricity, a flame, or a chemical reaction, is composed of many different colors that leave their source in all directions, without synchronization. In contrast, the light from a laser is made of only one wavelength and therefore of one pure color, and all the photons leave their source in a synchronized manner, traveling in a narrow beam. This type of light is called coherent.

Laser light is produced by atoms in the same way as ordinary light, but the atoms of the lasing material have a unique excited state that is very long-lived compared to the normal excited states. When the electrons of an excited atom reach this long-lived state on their way to the ground state, they can remain there up to a thousand times as long as in any other excited state. This long-lived excited state is known as a metastable state. Under normal conditions most electrons in atoms are in their ground state. However, when atoms with a metastable state are energized, more electrons can be found at the metastable state than in the ground state. This somewhat unnatural preference of "habitation" is known as population inversion, a condition critical for the production of laser light.

In 1916 Einstein had proposed the theory of stimulated emission. This idea is the heart of the mechanism of the laser, and Einstein is now regarded as its theoretical father. Einstein asked a very simple question: "What would happen if an electron in an excited state were to be hit by a photon of exactly the same energy?" After many calculations, he predicted that the photon would stimulate the de-excitation of the electron so that finally two photons of identical energy and color would "wiggle" along in a synchronized fashion and in the same direction. If on their way they were to encounter similar excited electrons, the process would be repeated, adding more photons to the original beam.

In ordinary atoms, where electrons stay only for a short time at an excited state, it is very unlikely that any of them will be "waiting" to be stimulated by an incoming photon. If, on the other hand, the atoms have a metastable state of the same energy as that of the incoming photons, then a large number of excited electrons can contribute photons by stimulated emission. One photon becomes two as it moves along, those two become four, and so on. As a result of this amplification, a single photon can become a powerful beam of laser light. This crucial mechanism is what gives the laser its name: Light Amplification by Stimulated Emission of Radiation.

The following diagram shows the mechanism involved in the production of laser light.

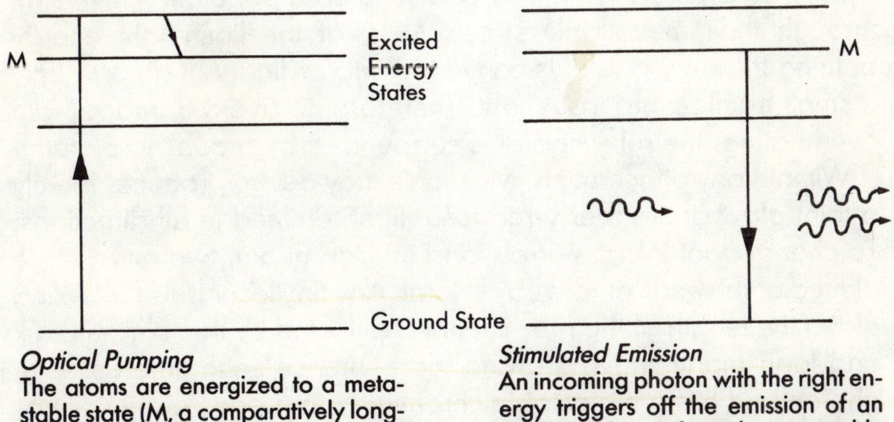

Optical Pumping
The atoms are energized to a metastable state (M, a comparatively long-lived excited state) to form a population inversion.

Stimulated Emission
An incoming photon with the right energy triggers off the emission of an identical photon from the metastable state (M). Both move on to trigger off more and more photons, so amplifying the original beam.

The first device that amplified radiation by stimulated emission was built by Charles H. Townes (1915–) in 1954.

Townes was aware of Einstein's theory on amplification by stimulated emission. His own expertise in radar and spectra led him eventually to the development of a historic device, the maser. This device contained ammonia, its metastable states used to amplify radiation of wavelength at 1.25 centimeters, which is in the microwave category (the *m* in maser stands for microwave). But these were no ordinary microwaves. They were of high purity of wavelength, bearing all the properties that were predicted for laser light.

The success of the maser encouraged other scientists to turn their energies to the development of a similar device that could produce visible light. In 1958 Townes and his Canadian brother-in-law, Arthur L. Schawlow (1921–), wrote a paper that described the theory and the methods by which laser light could be produced. Two years later, in July 1960, the first man-made laser light was produced by Theodore H. Maiman (1927–) at the Hughes Aircraft Laboratory in California.

The lasing material used to produce the first laser was synthetic ruby, an impure aluminum oxide with chromium atoms, which give ruby its characteristic red color. Maiman used a cylindrical ruby rod four centimeters long, with a flash tube coiled around it for bathing the crystal with light. When chromium atoms are energized by intense ordinary light, they can develop a population inversion through their metastable states. Most of the flash-light energy bathing the ruby crystal is converted into ordinary heat, and only a small fraction into laser light. Therefore, to avoid damage from overheating, the ruby crystal is surrounded by a cooling system.

Without any attachments, Maiman's ruby device produces merely a faint glow of cranberry-red laser light, emitted in all directions. To change that into a narrow and intense beam, two mirrors are placed at the ends of the ruby crystal. Any tiny laser pulse traveling along the length of the ruby crystal is reflected by the mirrors back and forth many times, growing more intense each time as each photon that hits a metastable chromium atom adds another photon to the beam. At some point, when the beam has grown to a very high intensity, it shoots right through one of the mirrors, which on purpose was made partially transparent. In Maiman's first ruby

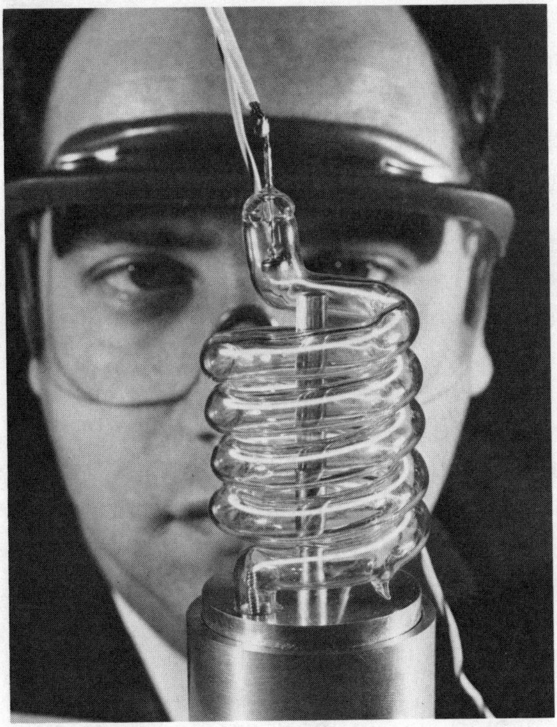

Dr. Theodore H. Maiman of Hughes Research Laboratories studies the main parts of his historic invention, the first working laser. The heart of this laser is a small rod of synthetic ruby crystal whose atoms, excited by the light of the coiled flash tube, generate laser pulses. (Courtesy Hughes Aircraft Company)

laser the outcome was a red laser pulse that lasted only 300 millionths of a second. This tiny pulse of pure light signaled the beginning of a new era in science and technology, one destined to transform the human landscape more than any other single invention.

Laser light is superior to ordinary light in many ways. It is pure, intense, and directional, qualities that put the "magic" in laser light. In a laser beam, the photons are of one color, and they all move in unison (in phase), reinforcing each other and so endowing the beam with spectacular effects even if its overall energy is small. For example, the energy of the pulse from the ruby laser is not enough to even boil an egg. Yet it can blast a hole in a three-millimeter-thick steel plate. How can this be? The answer is that blasting the hole depends not so much on the total energy as on how quickly this energy is delivered.

As an analogy, imagine a crowd throwing stones at the door of a castle to force it open. After a day of stone throwing, the door still stands, with nothing but scratches and with a large pile of stones at its base. Imagine that pile now as one rock, thrown at the door with the same energy as all the stones previously. The door will be smashed open. This is exactly how a laser pulse behaves.

The critical number is therefore not so much the energy of light as its power, a scientific term for energy delivered per second. If one joule (a unit of energy) is delivered in one second, then the power is one watt. But if one joule is delivered in one tenth of a second, then the power becomes 10 watts. A typical light bulb of 100 watts converts 100 joules of electricity per second into heat and light. If all that energy were delivered in a thousandth of a second, the power would be 100,000 watts.

The high degree of purity of laser light makes it ideal for many uses. In a sense, laser light is to ordinary light what a pure musical note is to noise. The purity of color enables laser light to identify the presence of tiny amounts of chemicals in liquids or gases. Chemical research, detecting atmospheric pollutants, analyzing blood, and even surgery are but a few of the applications based on this property. The nonspreading of the laser beam, its high degree of directionality, makes it ideal for uses in surveying and telecommunications.

Since 1960 the family of lasers has grown dramatically. They come in a variety of shapes and sizes, with different power outputs and a multitude of wavelengths. These can range from invisible—infrared, ultraviolet, and even X-rays—through all the colors of the rainbow. The type of wavelength emitted depends, of course, on the lasing material. In theory, any solid, liquid, or gas with a metastable state in its energy structure can produce laser light. A helium-neon mixture and carbon dioxide are the working substance of the two most popular gas lasers. Glass and semiconductors are the most popular materials of solid lasers. The list of lasing materials grows continuously, adding excitement to laser physics. Even food coloring has been used to make a laser. Some lasers, the tunable type, can produce any one of an infinite number of wavelengths with the use of fluorescent dyes. Each wavelength serves as a tool in a different task.

The method used to energize the lasing material differs from one type of laser to another. There are three major methods: the use of a high-intensity light (like the flash tube used by Maiman) or electrons from an electric charge (as in a fluorescent tube) or, finally but rarely, nuclear radiation.

In size, lasers can be as small as a grain of salt, with a power output of only a millionth watt. Others can fill a building the size of an airplane hangar, firing their powerful beams into minuscule pellets of nuclear-fusion fuel to produce energy. The power output of such lasers can be as high as 12 trillion watts, about twenty times as great as the power of all power plants in the United States. Again, it is not so much the energy as the rate at which it is delivered that makes these lasers so powerful.

Lasers can be classified in terms of size, intensity, wavelength, method of energizing, and type of material used. By the last method, all lasers fall into the category of solid, liquid, or gas.

Of the solid type, the most widely known are the ruby and the YAG laser. Both use light from powerful flash tubes to energize their atoms. Both use materials with host atoms in which the active atoms with the metastable state are imbedded. In the ruby laser the host atoms are aluminum oxide and the active ones are chromium. In the YAG laser active neodymium atoms are imbedded in a crystal made of yttrium aluminum and garnet—hence the name YAG.

The crystal of the ruby laser gets hot easily. The need for cooling prohibits its continuous use; it can operate only in pulses. The YAG crystal is a better carrier of heat and cools much faster on its own. This allows the YAG laser to operate in a continuous beam. If, instead of YAG materials, neodymium is mixed with glass, a new laser is produced—the glass laser. It is both cheaper and easier than the YAG laser, although not as efficient, because glass is a poor conductor of heat. Other crystals have been used in the manufacture of solid lasers. The ruby and YAG continue to be the most widely used.

Energizing the atoms of gas lasers is done by an electric discharge in which electrons flowing through the gas between two electrodes transfer some of their energy to the gas atoms. In the most commonly used model, the helium-neon laser, the flowing electrons energize the helium atoms, which then transfer their energy to the neon atoms that produce the laser beam. Although of a minor power output (less than one watt), helium-neon lasers are found in every science laboratory and are used in fascinating physics demonstrations. They are also used in surveying and in supermarket cash registers, and are planned for use in telecommunications.

There are at least five thousand varieties of metastable states in as many gases that can produce laser light of one wavelength or another. Lasers made with argon gas are very powerful and are widely used in industry. Krypton or krypton-argon lasers can produce a variety of brilliant color beams and are often utilized in spectacular laser shows.

The most popular of the industrial gas lasers is the carbon dioxide type. Its power output can be from as little as one watt to as much as hundreds of thousands. Such lasers are highly efficient and can deliver either a pulse or a continuous beam. The carbon dioxide laser is so widely used in industrial tasks, like welding, drilling, and cutting, that it is commonly referred to as the "workhorse" of industry.

There are two ways to energize the molecules of a carbon dioxide laser. If the molecules are forced to be excited to their metastable states, then the emitted light is near the visible range. If, on the other hand, the molecules are forced to vibrate, then their vibrational de-excitation produces an invisible beam that falls in the infrared category. Infrared laser beams have the ability to penetrate the atmosphere, which makes them suitable for weather research and possibly one day interplanetary communications.

In some gas lasers the excitation of the atoms is produced by the energy released in mixing two liquids or gases. Sometimes the combination requires a flame. For example, in the hydrogen-fluoride laser, hydrogen and fluoride gases flow into the laser chamber, where they are ignited to produce a flame that energizes the lasing atoms. Such lasers can produce very powerful infrared beams.

The excimer laser, designed in the mid-1970s, is a gas laser with a very unusual way of excitation that combines the chemical and electric discharge methods. Energetic electrons from an electric discharge cause a noble gas (argon, krypton, or xenon) to react with another gas (chlorine, fluorine, or iodine). The result is a molecule that exists only in an electronically excited state, known as

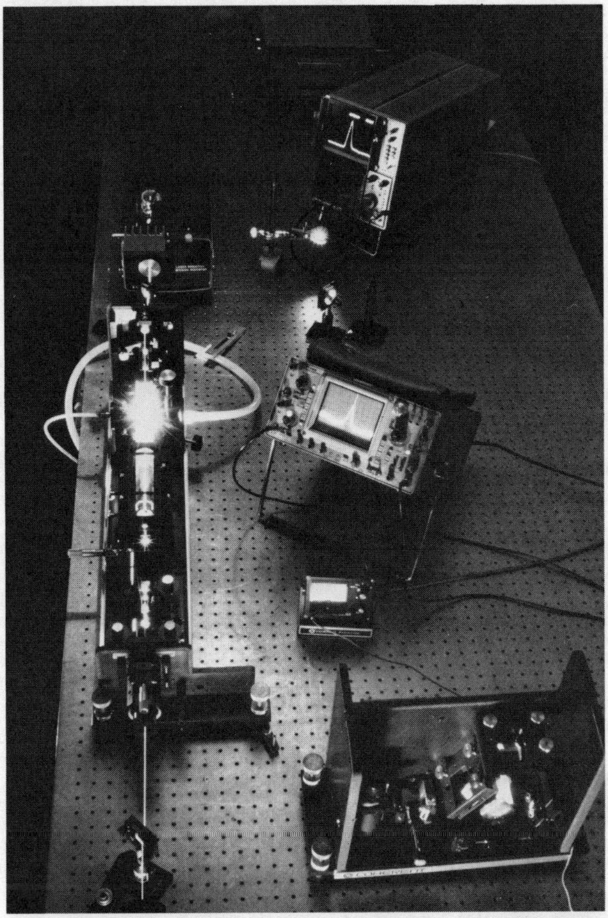

A Coherent dye laser performing a typical laboratory experiment. (Courtesy Coherent, Inc., Palo Alto, California)

the excimer. These laser beams are in the ultraviolet region, and they are used in chemical research and in the manufacture of semiconductors.

The family of lasers continues to multiply every year. More powerful, of new wavelengths with a different way of excitation and using different lasing materials, new exotic lasers provide for every need and pocket. Some are energized by nuclear fuel, generating laser beams of massive power. Some liquid lasers use dyes to produce beams that can be fine-tuned to any wavelength. Tiny and inexpensive red glowing semiconductor lasers, some the size of a grain of salt, are projected to have a working life of up to a hundred years. Further research in new materials may one day allow these minuscule lasers to have a working life of thousands of years. Their most effective use at the moment is in communications, sending pulses through glass fibers, vital nerves of the future network of telephone communications.

There is both mystery and excitement in the development of a laser that can emit powerful X-rays. Because of its potential military use in antiballistic defense, most of this work still remains a secret. There are two major problems in making such a laser: the energy required to excite atoms enough to produce X-rays is so great that it would vaporize the laser itself, and no mirrors can withstand such a powerful beam, so amplification poses a great deal of technological difficulty.

Work at the Lawrence Livermore National Laboratory suggests that a tiny nuclear explosion can do the trick. The ordinary X-rays produced in such an explosion can energize the atoms of the lasing material to produce laser-type X-rays. The total X-ray laser output is small, but since the output comes in a burst lasting only one trillionth of a second, its power can be hundreds of trillions of watts. Such a laser fired from an orbiting satellite could incapac-

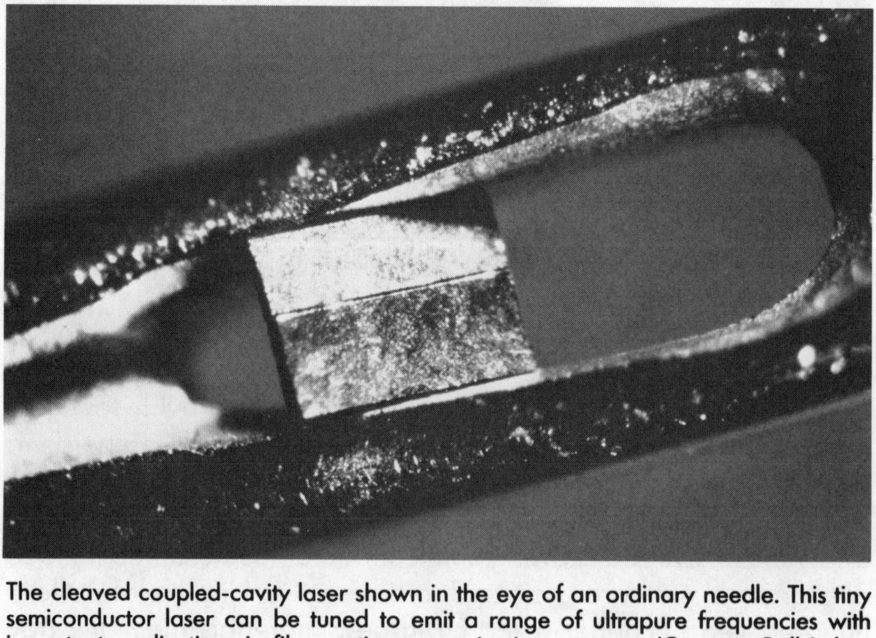

The cleaved coupled-cavity laser shown in the eye of an ordinary needle. This tiny semiconductor laser can be tuned to emit a range of ultrapure frequencies with important applications in fiber optic communications systems. (Courtesy Bell Laboratories)

itate an intercontinental missile (ICBM) minutes after takeoff. This is the heart of the proposed "Star Wars" defense program, whose projected cost and effectiveness have caused a lot of controversy among politicians and scientists alike.

Yet despite all that has been accomplished, this is still the neolithic age of the evolution of lasers. More efficient, versatile, and inexpensive types await their birth in research laboratories throughout the world. Better designs, longer life, new materials, better ways of energizing them, and greater precision and directionality are all bound to push the laser into a new era.

Laser Applications

Shortly after its birth the laser was called "a solution in search of problems." Scientists realized from the beginning that, just like electricity, this unique light was destined to become the most important single technological element of this era. After only twenty years of crude applications, the laser has already made an impressive impact on our way of life. It reads the price of groceries at the supermarket, translates the grooves of video disks into sounds and pictures, and transmits telephone conversations through glass fiber optics. Today the laser can perform technological miracles that were inconceivable only few years ago. From industry to medicine, from telecommunications to the war machine—one after another, sciences and technologies are turning to the laser for answers to their problems.

The laser owes its versatility to the unique characteristics of its pure light. When tuned to a particular color, it selects which target molecules to interact with. It can burn away a tumor or tooth decay, yet leave surrounding healthy tissue unaffected. Selectivity allows the laser to detect minuscule quantities of chemicals in fluids or pollutants in air. The power of a laser can be as small as a tiny pulse that delicately cleans masterpieces of art or as large as a titanic blast that triggers off nuclear fusion. Its great capacity to carry information enables a single beam to carry thousands of telephone lines or television channels. Its ability to pick up detail enables it to create magical 3-D photographs, known as holograms.

Excess material, vaporized by a laser beam, scatters in a rooster-tail pattern away from the surface of a batch of chips during a laser-scribing operation. (Courtesy Bell Laboratories)

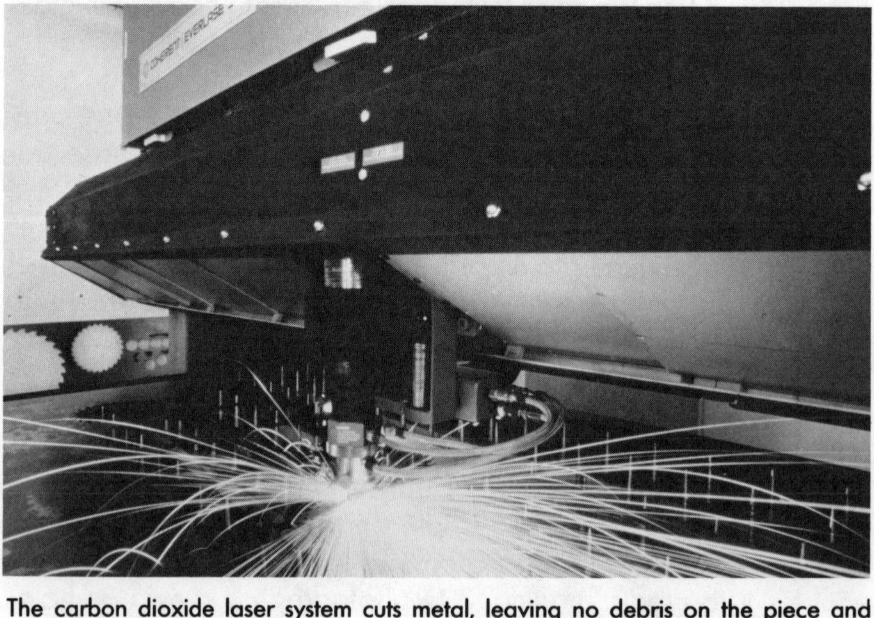

The carbon dioxide laser system cuts metal, leaving no debris on the piece and requiring no secondary finishing. (Courtesy Coherent, Inc., Palo Alto, California)

Some of the earliest applications of the laser were in industry. It can drill, cut, weld, and heat-treat better, more quickly, more accurately, and more efficiently than any conventional machine tool. Unlike mechanical tools, the laser never touches the object it treats. It merely shines light on it. This causes no vibrations and therefore ensures more accuracy. When focused to a fine spot, a laser beam can raise the temperature of target molecules by 100,000°C in less than one second, vaporizing the material in miniature explosions. Reduced to a superfine beam, the laser can etch the smallest print or draw tiny circuits for computer microchips.

In heavy industry lasers drill holes in diamonds, the hardest substance known, which are used as dies for drawing metal into superfine wires that are used in electronics. The conventional method of grinding the hole with diamond dust takes two to three days. With a pulse laser, it takes only two minutes. Fragile materials like ceramics that are likely to break under a metal drill can be

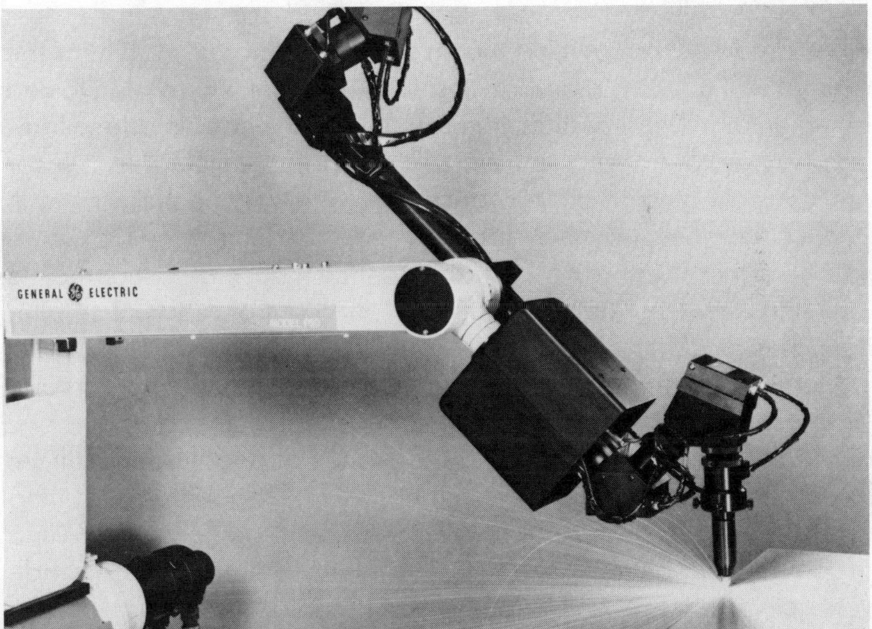

The marriage between robots and lasers is making a revolution in industry, particularly the assembly line. (Courtesy Coherent, Inc., Palo Alto, California)

drilled safely and quickly with a laser. Lasers are also used to cut saw blades out of a steel plate in just a few minutes, compared to hours with conventional cutting tools. In industrial robots, lasers controlled by computers perform dangerous, difficult, or monotonous tasks, like welding the frames of cars on an assembly line. Lasers heat-treat metals to produce extra-hard surfaces, essential for the many moving parts in engines.

Performing somewhat simpler tasks, lasers are already participating in the manufacture of common household items. They make the holes in baby bottle nipples and aerosol cans, and even the invisible tiny holes in cigarette papers that control the air flow, which in turn determines the amount of smoke intake. Lasers cut rubber, plastic, and even cloth. Cloth manufacturers already use computers that, fed with pattern information, guide a laser beam to burn-cut hundreds of suits per minute.

Despite their impressive success, lasers do have their limitations and drawbacks in industrial applications. Attempts to use them to cut food or plywood have failed. The food can be cut, but it is also burned. With plywood the laser is no economic match for the mechanical saw. In metal drilling a lot of laser energy is wasted by reflection. The lens that focuses the laser beam is often damaged by metal fragments from drilling. Holes drilled with a laser are conical rather than cylindrical, as with the mechanical drill. Because the laser has to be focused, there is a limit to the maximum size hole that it can drill. Last, the cost of industrial lasers makes them high-investment machine tools, often out of reach of smaller industries.

Medicine is another area in which lasers are making a major contribution. They unblock obstructions, heal ulcers, burn tumors, drill teeth, and restore eyesight. Thousands of people benefit from laser surgery each year in operations that no longer require cutting open the human body. The healing pulses are guided to the ailing organ through an optical fiber. Because laser light can be made very selective as to which specific tissue to vaporize and which to leave intact, the surgeon can operate precisely. This dramatically reduces pain.

Guided by thin, flexible fiber optics, which can be inserted through any of the natural openings of the human body, lasers can heal in a variety of ways. They destroy tumors in the colon and seal ulcers in the stomach. They can perform microscopic surgery on the larynx, vaporizing cancerous lesions without damaging the rest of the voice box. Laser surgery can also restore hearing impairments with delicacy and precision. The laser can chisel away bony substances in the inner ear without pain or bleeding. Like steam flowing over ice, the beam vaporizes excess bony substances that prevent proper movement in the bones of the inner ear, which is necessary for sound amplification. Through a fiber optic inserted through the vagina, laser light can unblock Fallopian tubes, restoring happiness through the gift of motherhood to thousands of women each year. In a similar technique, women with cancer of the cervix have been spared the pain of the scalpel. Most such surgery is now much cheaper and is performed on an outpatient basis.

Because lasers cauterize as they cut, they seal blood vessels and so prevent bleeding and reduce the danger of infection. Pulse lasers are used to remove warts, birthmarks, skin cancers, and even tattoos. In dentistry lasers drill or burn away tooth decay without pain. There are three reasons for that. Unlike mechanical drills, they produce no vibrations. Likewise, there is no heat developed by friction. Moreover, a dental-drill laser can vaporize only the dark decayed parts of a tooth, leaving unaffected healthy tissue, which is filled with nerves.

But perhaps the most widespread medical use of lasers is in eye surgery. Laser light, able to pass through the lens without injuring it, is used to cure a variety of eye problems. It can weld a detached retina, seal leaking blood vessels, break up microscopic particles floating in the gel-like fluid of the eye chamber, and, of course, destroy tumors. All these delicate microsurgeries are done on an outpatient basis, they produce little pain, and they cost less than the conventional scalpel methods. Additional uses of lasers in eye surgery are discussed in chapter 5.

Lasers are destined to play an ever-increasing role in medicine. One day X-ray lasers may enable molecular biologists to study the complete life cycle of a virus, enzyme, or even DNA molecule in extreme detail. Such information would undoubtedly unveil the mysteries of disease development, heredity, drug effectiveness,

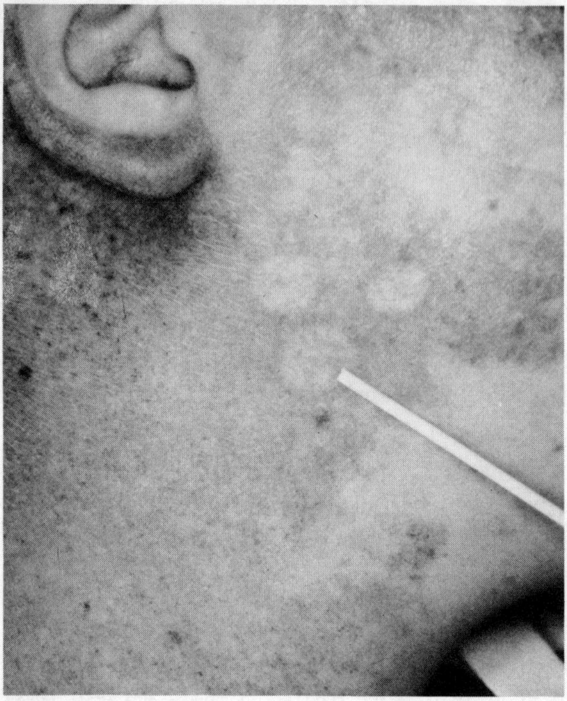

The bleaching effect of the argon laser on facial birthmarks. (Courtesy Dr. Leon Goldman, Director of Laser Treatment Center, Cincinnati)

immunity, and even aging. The laser's superior "sniffing" ability to identify "a needle in a haystack" may one day provide doctors with the perfect diagnosis. Lasers will be able to analyze a variety of human fluids, from blood to tears, seeking the presence of microscopic concentrations of illness-producing chemicals or cancer cells. Early treatment based on such an accurate diagnosis can arrest most medical disorders, assuring a healthier life.

The marriage of lasers with fiber optics is making an even greater impact in the field of telecommunications. It was in 1880 that Alexander Graham Bell (1847–1922), the father of the telephone, disclosed what he considered his most important invention—the photophone. In this brilliant but technologically premature gadget, Bell tried to use a beam of light as a carrier of voice. If successful, the photophone was to have replaced the telephone, which transmits sound as electrical signals through copper wires. Although the general principle is correct, Bell's photophone was a failure, mainly because ordinary light is obstructed by clouds, rain, and fog. After the discovery of the laser, the idea of light communication was resurrected, and this time with impressive results.

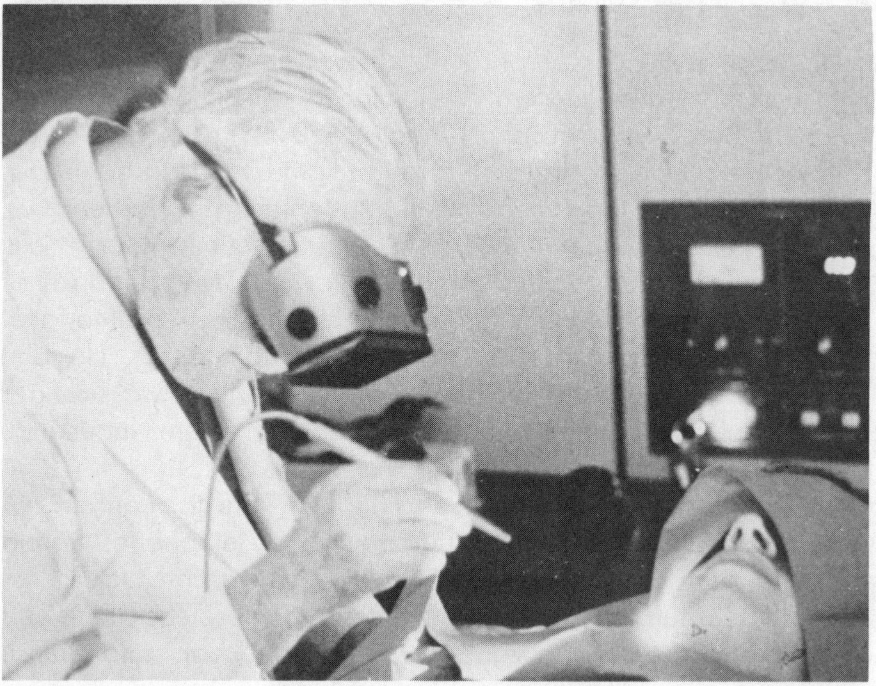

Laser surgery for the removal of marks on the face. The use of the argon laser requires eye protection. (Courtesy Dr. Leon Goldman, Director of Laser Treatment Center, Cincinnati)

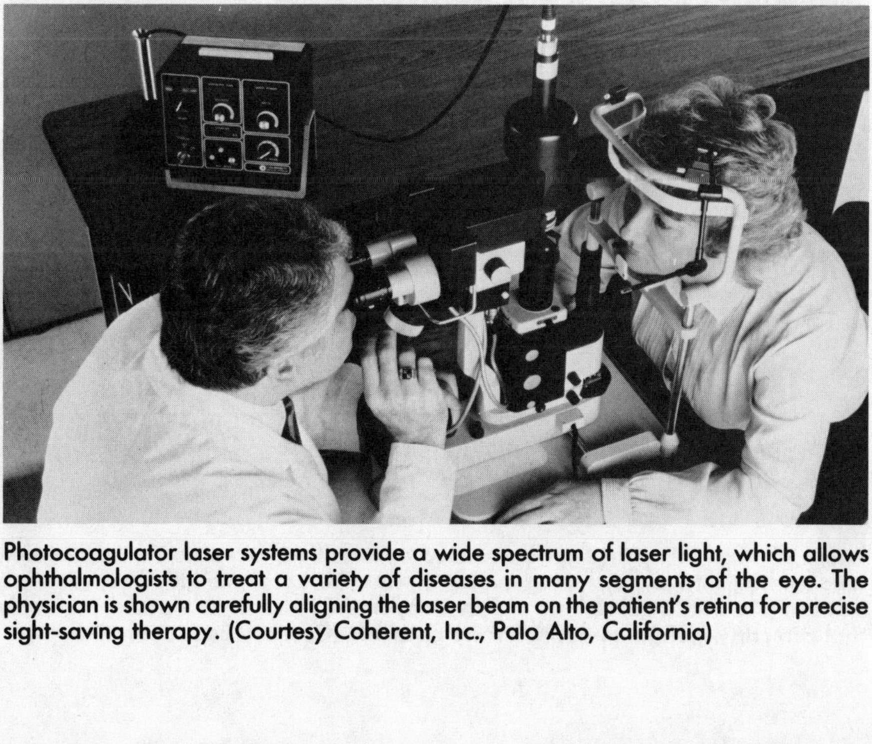

Photocoagulator laser systems provide a wide spectrum of laser light, which allows ophthalmologists to treat a variety of diseases in many segments of the eye. The physician is shown carefully aligning the laser beam on the patient's retina for precise sight-saving therapy. (Courtesy Coherent, Inc., Palo Alto, California)

Fiber optics are the second critical technology for a successful light communication system. Their birth is traced back to 1870, when the British scientist John Tyndall (1820–93) used a pipe with a reflective coating to transmit a ray of light by zigzag bouncing. If a glass rod is used instead, then light entering at one end will travel to the other, even if the rod is bent many times over. This is explained by the principle of internal reflection, by which a ray of light bounces back into the glass when it hits the glass-air interface. This is the same principle used by modern fiber optics. Flexible, high-purity glass fibers, thinner than a human hair yet stronger than a steel wire of similar diameter, are able to transmit thousands of telephone conversations on a ray of light.

There are two methods by which a beam of light can carry sound signals. The first is similar to the method used by telephones. Sound through the mouthpiece makes an electric current fluctuate in strength. Carried through copper wires to the earpiece at the end of the line, these electric signals are converted back into sound.

Loops of a hair-thin glass fiber, illuminated by laser light, represent the transmission medium for light-wave systems. Typically, twelve fibers are embedded between two strips of plastic in a flat ribbon, and as many as 12 ribbons are stacked in a cable that can carry more than 40,000 voice channels. (Courtesy Bell Laboratories)

A beam of light can be made to fluctuate in intensity in a similar way. But this method of intensity variation, known as analog, limits the amount of information that a beam of light can carry per second. If, on the other hand, the beam is spliced up into pulses like the dots and dashes in telegraphy, it can carry a billion times as much information per second as the analog method. Just as any written text can be reduced to dots and dashes in telegraphy, two single light pulses representing 0 and 1 can make up any sound signal, from a simple "hello" to a musical concert. They can also transmit pictures, making fiber optic cable television possible. The system that reduces all information to a combination of essentially only two signals or characters is called binary.

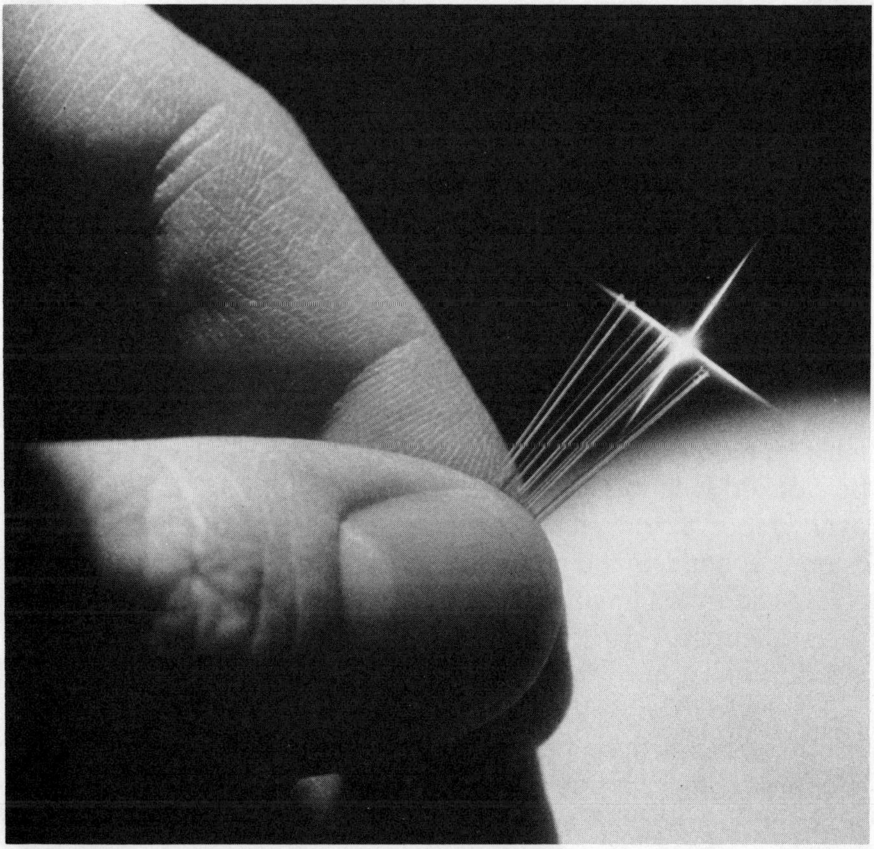

Hair-thin fiber optic light-guides that can carry voice and video data. (Courtesy Bell Laboratories)

Glass fibers that can carry voice, television, and data signals by a laser light. Such cables have already replaced conventional copper telephone wires. (Courtesy Bell Laboratories)

Although the use of laser light is desirable in fiber optic telecommunications, it is not really essential. Ordinary light from light-emitting diodes (LEDs) can also be used. Laser light, however, can be packed with lots more information, and it can be transmitted over longer distances. The amount of information carried by a beam depends on the switching speed of an electrically operated crystal. Scientists have managed to generate billions of pulses per second, enough to transmit fifty television channels or one hundred thousand telephone lines on a single beam.

Besides carrying more information, fiber optics provide better transmission, because they are less subject to electrical and weather interference than copper wires. They are more economical in cost, space, and weight. For example, a fiber optic cable the size of a finger has the same transmission power as eighteen conventional copper wire cables, each the size of a fist. This economy of space makes installation and maintenance much easier. Furthermore, fiber optics require fewer signal boosters or repeaters along their tracks. AT&T, which already has installed many miles of fiber optics, is planning a transatlantic fiber optic cable. Many firms in Japan, England, France, and the United States are investing billions of dollars to develop high-quality glass fiber and associated technologies that soon will make Bell's original idea of light communications a household reality.

The use of the laser has spread quickly to many other areas besides industry, medicine, and telecommunications. Combined

The several stages of glass as it is turned into optical fibers. Raw materials (the powdery substance at bottom) are melted to make disks of glass. In one version of the fabrication process, cylindrical blocks are formed into a rod and tube, and the combination is pulled into a fiber. Strands at top are finished fibers. (Courtesy Bell Laboratories)

with computers, lasers can eliminate long check-out lines at the supermarket. When the cashier moves a product over a tiny window, a low-power laser scans the coded black stripes found on each item. These dark lines make up the universal product code (UPC), which identify the nature and price of the item. The reflected laser beam brings this information to a detector, which converts it into electric signals that are sent to the store computer. In less than a second the computer subtracts that item from the store inventory and prints the price.

Lasers are used in modern cash registers to scan the price of groceries. (Courtesy National Cash Register Corporation)

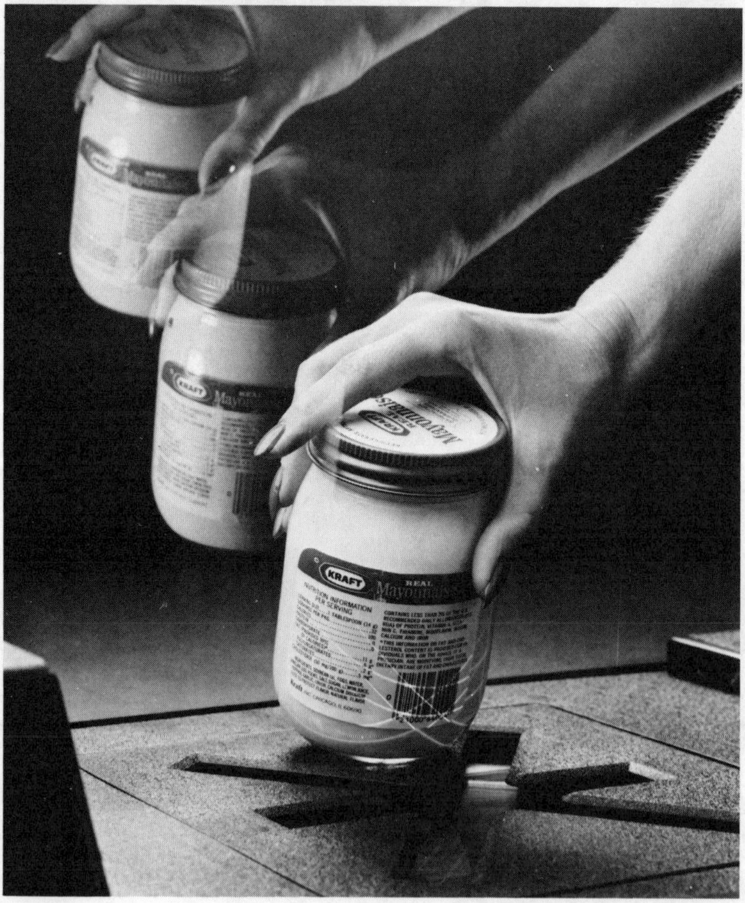

In the hands of surveyors, builders, and engineers, lasers have become the perfect measuring stick, the goniometer and the spirit level. They define boundaries and line up foundations, doors, and windows in the construction of buildings. As a goniometer they are used to grade the bottoms of pools, irrigation canals, and road surfaces for proper drainage. Surveying large areas with conventional tools used to take months. With the laser it takes only a few days.

Lasers of wavelengths that suffer little absorption by the atmosphere have great potential for satellite and interplanetary communication. They also perform continental surveying. When such a laser is reflected by a satellite from one continent to another, it allows scientists to measure the slow continental drift, which may be as little as one centimeter per year. Using the same technique, scientists can study the erratic motion of geological plates that cause earthquakes and volcanic activity. From the time it takes for a similar laser to bounce back from the surface of the moon, scientists can calculate the distance between the earth and the moon to an accuracy of six inches.

The use of the laser as radar is already making a contribution to understanding and forecasting the weather. The theory of operation is very simple. Analysis of a laser beam that bounces off an object yields information about the distance and nature of the object. The shorter the wavelength of the laser, the smaller the particles it can detect. Such use of lasers is known as LIDAR, from Light Detection and Ranging. In this way lasers can detect water vapor, dust, pollutants, atmospheric turbulence, and even temperature layers. Ground- or satellite-based, these weather lasers when connected to computers may one day be able to prepare an accurate long-term weather forecast, with dramatic benefits to industry and agriculture.

Laser applications have been few and rather slow in art. Lately, however, their use, although still limited, has been impressive in a variety of different areas. In laser shows, they draw giant multicolor diagrams on a night sky. The spectrum of a laser reflected from a painting can provide information about the chemical composition of the paint. Comparing that to facts such as that zinc was not used till 1820 helps uncover art forgeries. Another of the laser's qualities—that of a superfine energy control—helps art restorers vaporize thin layers of dirt or even secondary paint covering an original masterpiece.

The most exciting use of lasers in art is in the making of three-dimensional images, known as holograms. (The term was proposed by Dennis Gabor [1900–1979], the inventor of the hologram, and comes from the Greek words *holos*, whole, and *gramma*, message.) These ghostly-looking laser images are both beautiful and extremely useful in industry, computers, and medicine.

A hologram is made by exposing a photographic plate to laser light reflected from every part of an object. The original laser beam is first split up into two parts. One part, reflected only by a mirror, spreads directly onto the plate and is known as the reference beam. The other part bathes the object, and the reflections are directed to the plate. Thus, every single grain of the photographic plate receives the reference beam and an infinite number of other laser signals, each coming from a different part of the object. All these signals interact with the reference beam to produce a permanent and complex granular pattern.

When the hologram is illuminated by a laser (some need only ordinary light), different views of the image are seen from different angles. The result is a genuine 3-D image with depth and definition. If more than one color laser is used to make the hologram, the image is a brilliant multicolor 3-D ghostly delight. The fascinating thing about the hologram is that the entire image is captured in every single piece.

Beyond their artistic beauty, holograms were discovered to have a far greater potential in science. In medicine, holograms of a single cell or a virus in action can provide clues about disease and how best to combat it. In industry, engineering, and construction, holograms can detect tiny stress differences in a body, information that can improve the design, quality, and safety, and lower the cost, of many engineering products and construction projects. Their ability to identify even the tiniest of stress variations is mind-boggling. They can detect, for example, the exact location of a fly landing on a concrete slab. The computer industry is viewing holograms as the ultimate filing cabinet for storing almost infinite amounts of information. With the development of a proper material, computer scientists believe, one day a hologram the size of a sugar cube will be able to store the name, address, and telephone number of everyone in the world with a telephone.

As with most modern technologies, lasers are part of the military arsenal. Penetrating the depths of the ocean, laser beams are soon to become the ears of submarine-hunter planes. Their reflections from an enemy tank guide a missile to a bull's-eye hit. Thanks to their great speed, they can deliver a destructive pulse instantly to anything in sight. Combined with satellites and computers, they may one day fight a space war. Fired from satellites or from the ground, they have become the key element in a controversial proposal for a multiple line of beam defense against intercontinental missiles. Some feel it will bring an end to the horrible prospect of a nuclear holocaust. To others it is a wasteful system of defense despite its impressive "Star Wars" outlook. It is at least very fortunate that, unlike the atom bomb, this other brainchild of Einstein remains by and large a defensive weapon.

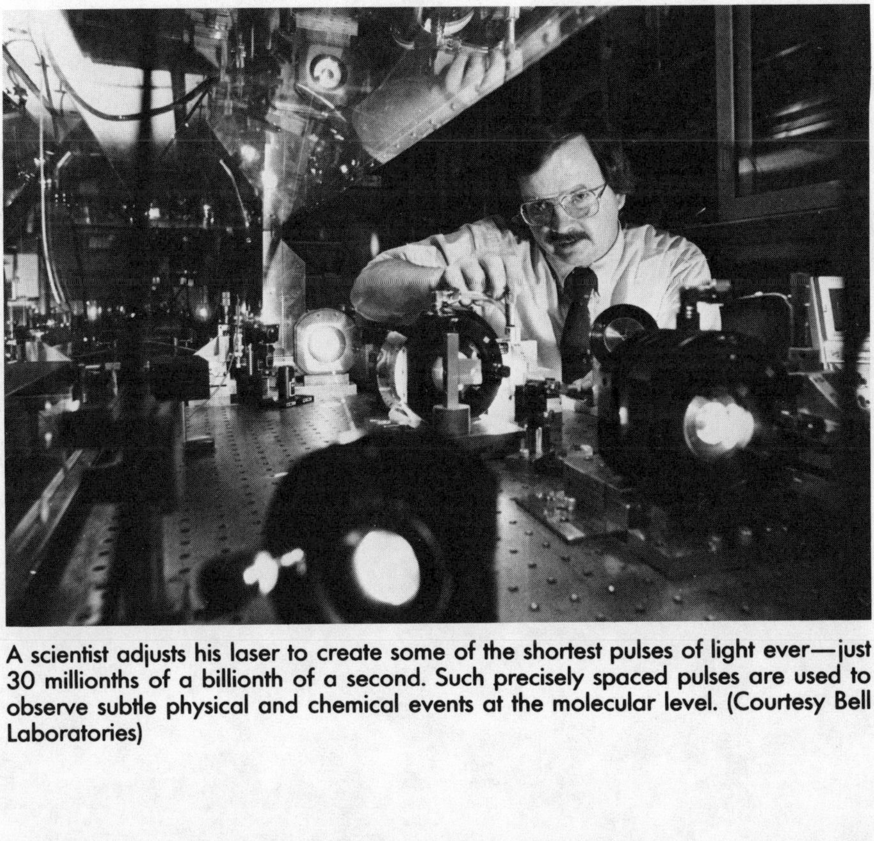

A scientist adjusts his laser to create some of the shortest pulses of light ever—just 30 millionths of a billionth of a second. Such precisely spaced pulses are used to observe subtle physical and chemical events at the molecular level. (Courtesy Bell Laboratories)

If not our savior from a nuclear war, the laser is certainly rallying to our rescue from another problem: the energy crisis. There is a lot of optimism that by 2020 giant lasers will produce low-cost nuclear fusion—the process that powers the stars. Although the theory is well understood, the technological difficulties are still formidable. When solved, nuclear fusion will bring about monumental benefits. For example, fusing the hydrogen in a bucket of water could generate enough electricity to power a city, with no waste to store or to pollute the environment. Nuclear fusion is discussed in detail in the next chapter.

From performing delicate eye surgery to blasting ICBMs, from transmitting pictures and sound to holography, the laser exhibits an amazing versatility to suit many of our needs. Many believe that the best is still to come. When married to other technologies, such as satellites and computers, the laser opens up new horizons and many more possibilities. Any one of its main applications would have been enough to place it in the category of great innovations.

The ability to communicate is the single ingredient that is absolutely essential to the advancement and prosperity of any civilization. The inventions of writing, paper, printing, the telephone, the radio, and now the laser and the computer—all stand out as the very key to a new age in human progress. The laser and the computer have already launched a new information revolution unparalleled in history.

Chapter Three
Nuclear Energy

As nature's eternal soul and as the driving force of the universe, energy is everywhere, manifesting itself in many different ways. It appears as light, heat, and sound; is in the motion of every moving object; and is stored in the chemical makeup of matter. In the nineteenth century scientists discovered that energy obeys a conservation law and that, in essence, it is indestructible. In 1905 Einstein proposed that energy and matter are manifestations of the same thing, that matter is really concentrated energy. This led to the development of nuclear energy—a new and more powerful form of energy derived from within the nucleus itself.

The utilization of energy was a key to the greatest periods of human progress. In its earliest use as fire, energy protected, warmed, cooked, illuminated, baked clay pots, and smelted metals for tools and weapons for our early ancestors. Raw muscular energy from domesticated animals enabled people to develop agriculture, ride to battle, and expand trade routes. With wind energy, ships sailed across seas and oceans, and the versatile use of water power triggered off a mini–industrial revolution in the Middle Ages. The major industrial revolution that followed in the eighteenth and nineteenth centuries drew its steam power from coal.

Post–World War II industrial expansion and population growth increased our appetite for energy dramatically. Escalating oil prices and the Arab oil embargo of the 1970s underline the possibility of an energy crisis and the urgent need to conserve and to look for alternative sources of energy. Nuclear energy was seen earlier by many as a viable alternative for cheap and abundant energy. Despite the successful performance of many nuclear reactors in this country and abroad, public concern about the dangers of radioactive leakage, reactor accidents, and the overall escalating cost of reactor construction has made the nuclear program very controversial indeed. Optimists assure us that the near future will bring a solution to most of these problems, as well as a breakthrough in a new form of nuclear energy—nuclear fusion—that promises inexhaustible amounts of cheap and clean energy.

The Historical Development of Nuclear Energy

On the afternoon of December 2, 1942, a dramatic experiment was about to reach its conclusion. Its outcome was destined to be possibly the most critical single discovery in the history of science, promising prosperity and progress as well as the threat of the ultimate destruction of all life on earth. In absolute secrecy, beneath the west stands of the University of Chicago athletic stadium, a group of forty-three European and American scientists was about to carry out the first self-sustaining nuclear chain reaction.

In the middle of the room was a large pile of graphite bricks with a few tons of enriched uranium oxide in its center. Sensors in the pile were wired to instruments a few feet away. A major control rod, nicknamed Zip, was gradually withdrawn from the pile. The scientists, their eyes glued to the instruments, were filled with mixed excitement and terror. They were stepping into uncharted territory. No one had tampered with nature's most intimate component, the nucleus, in such a fashion before. They were all aware that the experiment might possibly end in a devasting explosion. At 3:30 P.M. the team leader, the Italian Enrico Fermi, announced the success of the experiment by saying, "The reaction is self-sustaining. The curve is exponential."

After a simple celebration over a glass of wine, Arthur Compton, a team member, telephoned the results to James B. Conant of Harvard University. "The Italian navigator has landed in the New World," he said. Without a doubt, Fermi was the Prometheus of a new kind of energy that introduced the world to the atomic age of nuclear reactors as well as nuclear weapons.

Fermi's experiment was part of a larger project, code-named "Manhattan": the development of the first atomic bomb. The project began soon after President Roosevelt received a letter from Albert Einstein in August 1939. Einstein informed the President of the possibility of a superbomb based on the fission of uranium and recommended a cooperative effort between government and scientists to promote research and secure a supply of uranium. On August 6, 1945, after a two-billion-dollar secret budget had been spent on the Manhattan Project, the first atomic bomb was dropped on Hiroshima. A few days later, after a second atomic bomb was dropped on Nagasaki, Japan surrendered.

It is unfortunate that nuclear energy was born out of a military need and that humanity's first taste of it was through death and ruin. Whether it was appropriate to use the bomb when Germany had already surrendered and Japan was clearly losing continues to be an issue of heated debate. Many scientists, Einstein among them, feared an inevitable nuclear arms race and tried passionately to establish international agreements on the control of nuclear technology. J. Robert Oppenheimer (1904–67), the director of the Manhattan Project, acknowledged the tragic failure of the scientific community to stop the nuclear arms race and said years later that humanity had lost a golden opportunity for permanent world peace built on respect for and fear of the nuclear genie.

The Manhattan Project is a landmark not only in military history but also in the history of science. Never before had such a large group of scientists coordinated their research efforts for a single purpose. Never before had so much money been spent by the government on a scientific project. Never before had the military

The first atomic explosion, code-named "Trinity," was a plutonium fission device. It took place in the New Mexico desert on July 16, 1945. A few days later, the first atomic bomb was dropped on Hiroshima. (Courtesy American Institute of Physics, Niels Bohr Library)

establishment used the cream of scientific intellect to such a degree. The project was a turning point in the future of big science. Some scientists saw this as a backward step, in that science lost its innocence and independence, selling its spirit to the government. Others saw this as a beneficial relationship and an inevitable step in the evolution of science.

It took almost ten years after the A-bomb before controlled nuclear fission was put to practical use. In 1954 the first vehicle propelled by atomic power, the submarine *Nautilus*, was launched. In 1956 the first atomic reactor for the production of electricity, at Calder Hall in England, became operational. One year later the first American atomic electric power station, the Shipping Port Reactor, began its operation. But although the story of nuclear energy begins with the Manhattan Project, its prehistory goes back to the end of the nineteenth century.

In 1897, only one year after Roentgen discovered X-rays, Becquerel discovered radioactivity. This invisible and penetrating radiation that mysteriously fogged his film was studied thoroughly by the Curies and by Ernest Rutherford. Their work showed that there are three kinds of radioactivity, all originating in unstable nuclei, such as uranium and radium, which eject particles and excess energy in strong moves toward stability.

The puzzle of the source of this apparently inexhaustible energy was solved by Einstein. His formula $E = mc^2$ suggested that a small amount of mass is a huge concentration of energy. The production of nuclear energy is based on this formula. When uranium, a major fuel in nuclear reactors, is fissioned (its nucleus split into smaller fragments), some of the original mass becomes energy.

During the next forty years individual discoveries contributed to the overall theory required to achieve nuclear fission. Soon after discovering the nucleus in 1911, Rutherford achieved the first artificial transmutation of an element by nuclear bombardment. The experiment showed that by using appropriate particle bullets it is possible, like a twentieth-century alchemist, to transform one element into another. In 1932 James Chadwick (1891–1974) discovered the neutron, which can be the ideal particle bullet. It carries no electric charge, so it can hit the nucleus undeflected by any electric repulsion. The same year, Cockcroft and Walton verified experimentally Einstein's $E = mc^2$ formula.

In 1934 Enrico Fermi (1901–54) discovered that neutron bombardment could disintegrate many atoms, including those of uranium, with which he worked for many years. This was a key discovery, which he would use eight years later to build the first experimental atomic reactor, in Chicago. When, together with his family, he went to Sweden to receive the Nobel Prize in physics in 1938, the Italian fascist press blasted him with criticism for not wearing the fascist uniform and failing to give the fascist salute at the award ceremony. Seeing the dark cloud of fascism descending over Europe, Fermi moved to the United States, joining the many refugee scientists from Europe and, like many of them, dedicating his skills to the defeat of fascism.

In the fall of 1938 two German scientists, Otto Hahn (1879–1968) and Fritz Strassmann (1902–), discovered that after uranium was bombarded with neutrons, barium was mysteriously present in the residue. Since barium has a nucleus about half as massive as that of uranium, the only explanation of its origin was that it was the result of a uranium atom split, or fission. Lise Meitner (1878–1968) analyzed the results mathematically, and demonstrated that indeed this was the case. But she also found that the total mass of the residual atoms was less than the mass of the participating uranium atoms, as if in fact some matter disappeared during fission. Further calculations showed that the "missing" matter was transformed into energy, exactly as Einstein's formula suggested. The experiment of Hahn and Strassmann and the calculations of Meitner alarmed the scientific community of the free world. President Roosevelt had to be informed, and scientists chose Albert Einstein, their most famous colleague, to carry out this task. Einstein was someone to whom the President would definitely give his attention.

Nuclear Fission

Many heavy atoms have an unstable nucleus because it is packed with too many protons and neutrons. Such a nucleus tends spontaneously to emit energy and matter, seeking a more stable structure, which is a smaller and, of course, entirely different nucleus. This process is known as radioactive decay, and what is emitted is known as radioactivity. The time is takes for half the original nuclei to be transformed is known as half-life. Different radioactive substances have different half-lives, some as short as a few seconds, others as long as millions of years. The long half-life of some nuclear energy waste products has been a major concern in designing the best system of nuclear waste storage.

There are three kinds of radioactivity: alpha and beta particles and gamma radiation. Each has a different penetrating and ionizing power. Ionization is the removal of an electron from a molecule by collision—in this case, with radioactivity. This is radioactivity's most dangerous property, since ionization can affect the genetic structure of cells. Ionized cells can produce mutations, which when out of control result in cancer.

Of the three, the most ionizing are alpha particles, which are made of two protons and two neutrons. They are slow-moving, and they can be stopped easily—even by a piece of paper. They are extremely dangerous even in tiny quantities when taken in accidentally through breathing or through radioactively contaminated food or water. Beta particles are made of electrons, and they are the by-product of a transformation in which a neutron changes into a proton inside the nucleus. They are faster and more penetrating than alpha particles, but their ionizing power is a hundredth that of alpha particles. Gamma rays travel with the speed of light and are extremely penetrating, even more than X-rays. It takes about a meter of concrete or ten centimeters of lead to stop them. Fortunately, their ionizing power is small, a mere ten-thousandth that of alpha particles.

There are many ways to detect radioactivity. The most common is the use of the Geiger counter. This instrument converts ionization caused by radioactivity into clicks of sound and a meter reading. The stronger the radioactivity, the more ionization it can cause inside the counter, and the higher the produced signal.

Even when all radioactive sources are removed, the counter continues to click. What the counter is recording is background radiation caused by cosmic rays coming from space and by radioactivity occurring naturally in our environment. The earth's crust, mountains and lakes, and even houses continuously emit tiny amounts of radioactivity. Even the food that we eat and parts of our body contain minuscule amounts of radioactivity. We have indeed evolved in an ocean of background radiation.

The unit of radioactivity as it affects the biology of human cells is the millirem. (One rem—the term is shortened from "roentgen equivalent, man"—is the amount of ionizing radiation that produces a biological effect equal to that of one roentgen of X-rays or gamma rays. One mrem is a thousandth of a rem.) The typical average dose of radioactivity that we receive from background radiation (including X-rays) is about 200 mrem per year. The human body absorbs about 25 mrem in a chest X-ray, about 25 mrem per year from water, food, and air. A London–New York air round trip can add 4 mrem.

Radioactive materials, often referred to as radioisotopes, are by no means an inevitable evil of the nuclear age. With proper and careful use, these dangerous materials can make a valuable contribution to our lives in so many different ways, such as in industry, medicine, and agriculture. For example, gamma radiation from cobalt 60 is used to sterilize hypodermic needles after they have been sealed and packaged. Radioisotopes can act as tracers in fluids to determine the source of chemical pollution in rivers and lakes or the leakage of an underground pipe. Injected into the bloodstream, they can help identify clots, poor heart action, and other major organ malfunctions. In large doses, radioactivity can be focused on a single spot to destroy cancer cells without the need of surgery.

The use of radioactivity in food preservation may one day save billions of dollars in wasted agricultural products. Mild irradiation of grains and potatoes prevents them from sprouting. In a similar way, the irradiation of fresh fruits and vegetables kills microorganisms that cause rotting. Although this has already been tried on a small scale—in the space program, for example—a wide use of radioactivity in food preservation may be some time away. It will require strong evidence that can reassure the public of its safety.

Until 1972 everyone thought that the first time a self-sustained nuclear fission took place on earth was on December 2, 1942. However, geological evidence from an open-pit uranium mine in the Republic of Gabon, on the west coast of Africa, shows that nature beat us to it by at least two billion years. It so happened that all the conditions necessary for self-sustained nuclear fission came together naturally. What exactly are all these critical conditions that can make such a reaction possible?

Nuclear fission begins when a slow neutron collides with a uranium-235 nucleus. The collision causes the nucleus to split into two fragments of about equal size and two or three other neutrons. Under the right conditions these neutrons cause more fissioning, each time adding two or three neutrons to the stream of nuclear bullets. This avalanchelike increase of fissions is known as a chain reaction.

In each fission a small part of the nuclear mass is converted entirely into energy; the mass of the fragments after fission is always a little less than the nuclear mass before fission. The fragments travel at high speeds in all directions, carrying the energy released during fission as energy of motion, or kinetic energy. Although the energy released by the fission of a single nucleus is infinitesimal, the vast number of participating atoms can add up to enormous amounts of energy. To illustrate this point, consider a night-light that uses about 7 watts of electricity. It would require the fission of 210 billion uranium atoms per second to produce that electricity. But even a grain of salt contains many more atoms than that. If it were possible to convert one pound of matter entirely into energy, it would provide the electricity for an average home for 100,000 years, or drive a car around the world 180,000 times, or propel a large oil tanker around the world 400 times. But even in nuclear fission, only a tiny amount of the fuel is converted entirely into energy.

Neutrons and uranium-235 atoms are essential but not the only requirements for self-sustained nuclear fission. Two very important conditions must also be satisfied. The first one requires a minimum number of available target atoms. The smallest amount of uranium mass able to sustain nuclear fission is known as critical mass. Any amount less than that is like a fire without enough wood to build up to a self sustaining level. The second precondition is related to the speed of the neutrons that split the uranium atoms. Their speed must be about 1.2 miles per second. Like golf balls rolling toward a hole, they will never make it to the target if they are too slow, but if they are too fast they will not be captured. Since fission neutrons travel at very high speeds, a material known as a moderator is used to slow them down. The most efficient moderators are beryllium, graphite, and heavy water (water whose normal hydrogen is replaced by a heavier hydrogen atom known as deuterium).

A nuclear and a conventional power plant have very similar principles of operation. Both generate electricity by steam-driven turbines that spin electric generators. But whereas a conventional plant burns coal or oil to produce steam, a nuclear plant uses the heat generated by nuclear fission. Sustaining and controlling nuclear fission safely demands the construction of a complex and highly engineered superstructure, strong enough to withstand earthquakes or an airplane crash.

There are four basic components in a nuclear reactor: the fuel, the moderator, the control rods, and the cooling system that removes the heat generated by fission. Except for part of the cooling system, all of these components are housed inside the reactor vessel, a twelve-inch-thick, forty-foot-high carbon steel container lined with stainless steel. The reactor vessel is surrounded by a concrete wall as much as ten feet thick, and this in turn is covered by the containment building, a steel-lined reinforced concrete shell about two hundred feet high, with walls up to four feet thick.

High-temperature gas-cooled reactor (HTGR)

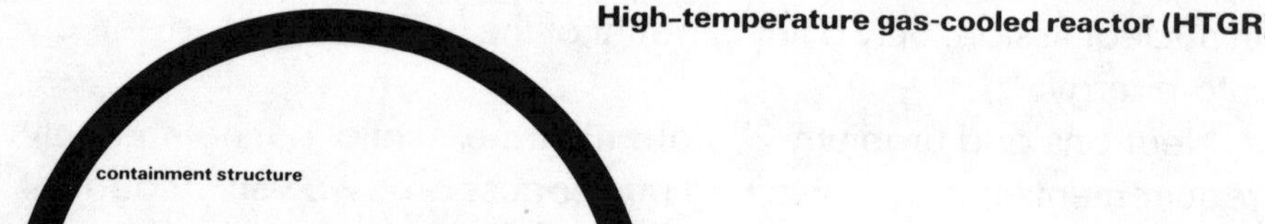

A high-temperature gas-cooled reactor. The reactor shown here is fueled by uranium carbide particles distributed in graphite in the core. Helium gas is used as a coolant to transfer the heat from the core to the steam generator. (Courtesy Atomic Industrial Forum, Inc.)

Boiling-water reactor (BWR)

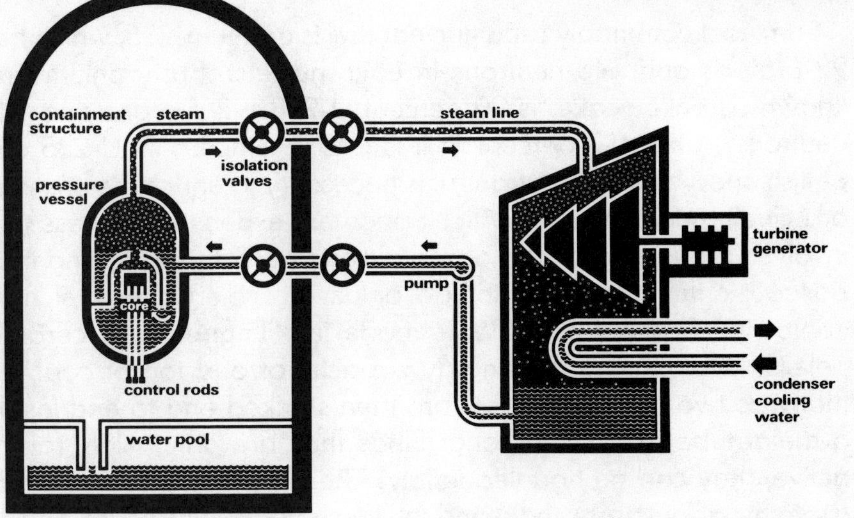

A boiling-water reactor. In this type of nuclear reactor, water is pumped through the core, boils, and produces steam, which is piped to the turbine. (Courtesy Atomic Industrial Forum, Inc.)

Pressurized-water reactor (PWR)

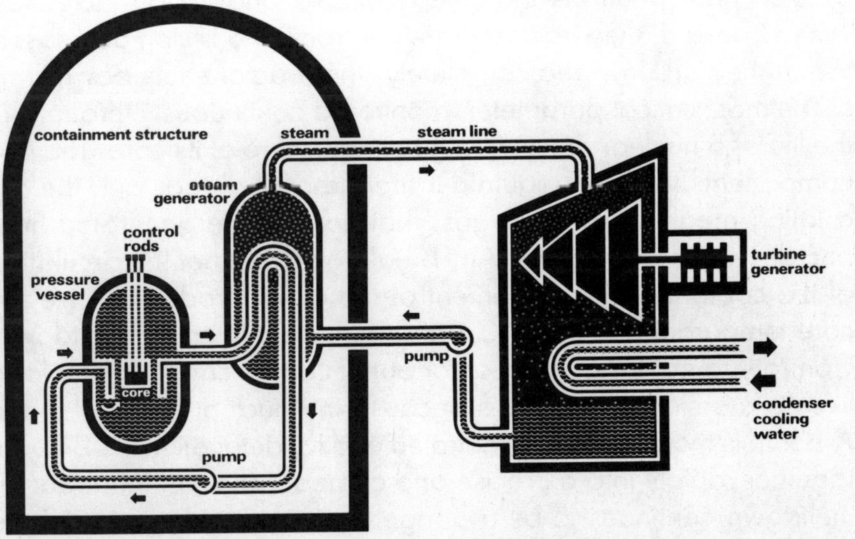

A pressurized-water reactor. Water under pressure, to prevent it from boiling, is circulated through the core to transfer heat to the steam generator. (Courtesy Atomic Industrial Forum, Inc.)

The most commonly used nuclear fuel is uranium-235, which has 92 protons and 143 neutrons in each nucleus. But uranium ore, known as yellowcake, is 90 percent U-238 (92 protons and 146 neutrons). Only 0.7 percent of it is U-235. Since only U-235 can be fissioned by slow neutrons, it is necessary to enrich the uranium ore. Fuel enrichment is a difficult and very expensive process that involves reducing uranium ore to a gas and then separating it by gaseous diffusion, gas centrifuge, or lasers. The enriched fuel, now containing 3 percent U-235, is made into fingertip-size ceramic pellets, each having the energy capacity of one ton of coal. As many as two hundred pellets are then stacked end to end inside a metal tube twelve feet long. Since they are only mildly radioactive, they can be handled safely. The core of a typical reactor contains about forty thousand of these rods, which are tied in bundles of two hundred and are known as fuel assemblies. A fuel assembly remains in the reactor up to four years. Spent fuel is replaced every year, however.

The control rods, made of a good neutron absorber such as cadmium, are inserted between the fuel rods, and their up-and-down movement controls the rate of fission. When they are raised a little, more neutrons become available and fission increases. When lowered, they absorb more neutrons and fission decreases. When they are lowered completely, the reactor shuts down.

The most critical parameter, monitored continuously throughout the life of a nuclear reactor, is the temperature of its core. The two components that can regulate it are the control rods and the circulating medium, liquid or gas, that collects the generated heat and delivers it to the boiler. Any breakdown in either the circulation of the coolant or the movement of the control rods will raise the core temperature to 3,000°C. This is sufficient to cause meltdown. Contrary to public fear, a reactor out of control can never explode like an atomic bomb. This is because the nuclear material in the A-bomb is more highly concentrated and for detonation it is brought together rapidly into a precise and compact shape. But a reactor meltdown continues to be a dangerous accident because of the potential leakage of radioactive materials into the soil, water, and atmosphere.

The number of nuclear reactors in operation or under construction is in the hundreds, distributed in at least twenty-five countries. Although they all share a common principle of operation, they differ in size, in design, and, most important, in their heat-recovery method. Most of the one hundred or so American reactors use circulating water at high pressure. Some others use circulating water that turns directly into steam. Canadian reactors use heavy water at ordinary pressures. The British use carbon dioxide gas at high pressures. The extremely high temperatures developed in the core of the breeder reactor, an experimental reactor that produces its own fuel, require the circulation of liquid sodium, which as metal can transfer a lot of heat without the need to be under pressure. The fact that liquid sodium reacts violently when it comes in contact

A technician checks new fuel assemblies prior to loading of a light-water reactor. Each assembly consists of approximately 250 rods packed with uranium pellets. A fuel assembly such as this will generate as much electricity as 100,000 barrels of oil. (Courtesy Atomic Industrial Forum, Inc.)

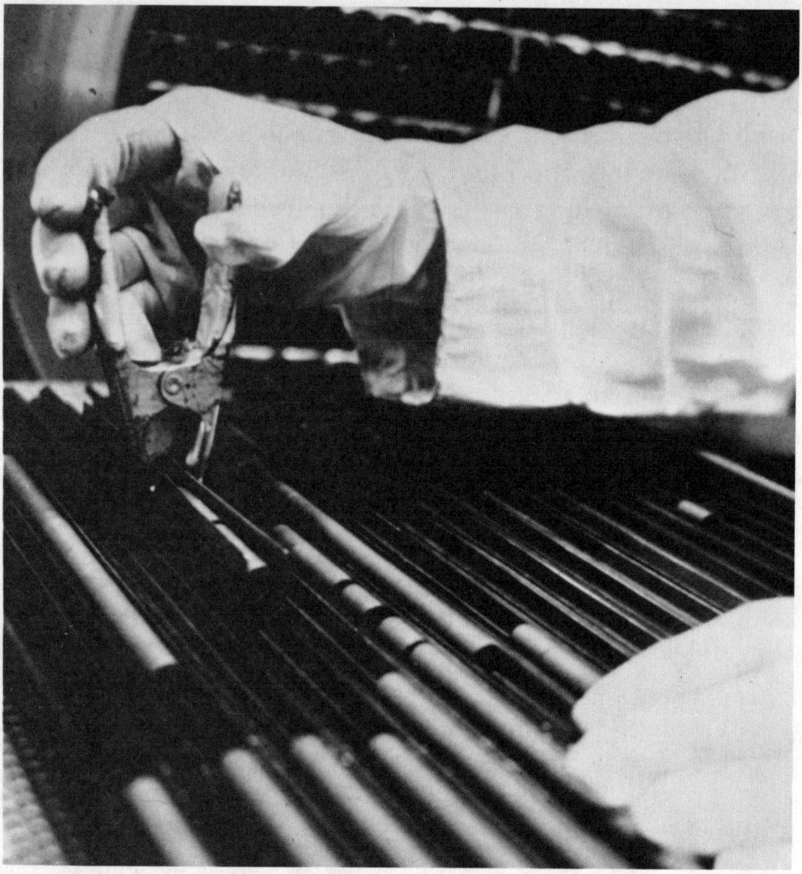

Uranium fuel fabrication. After uranium has been mined, milled, converted, and enriched, the uranium hexafluoride gas is processed into uranium oxide powder, which is then formed into half-inch-long pellets. The pellets are placed in thin metal (zirconium alloy) tubes twelve to fourteen feet long. Each fuel pellet can generate as much electricity as about 1,600 pounds of coal. The fuel rods are bundled together to form fuel assemblies. Several hundred assemblies form the reactor core of a large nuclear power plant. (Courtesy Atomic Industrial Forum, Inc.)

with water or steam imposes an extra burden on the safety design.

The breeder reactor is a mystery to many. Indeed, some may see it as a perpetual-motion machine violating the second law of thermodynamics. How does the breeder reactor produce more fuel than it consumes? Unlike uranium-235, uranium-238 cannot be used as a nuclear fuel, because it is not fissionable. But when exposed to neutron bombardment inside the breeder reactor, it is transformed into a new element, plutonium-239 (U-238 with an extra neutron in its nucleus). Plutonium is a fissionable material

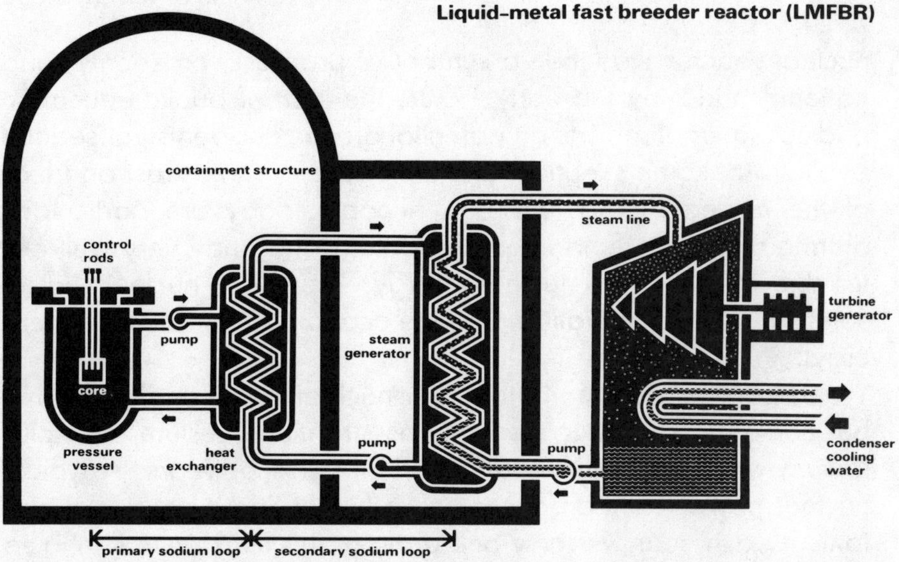

A liquid-metal fast breeder reactor. Molten sodium in the primary loop is pumped through the reactor core, containing the fuel. It collects the heat and transfers it to a secondary sodium loop in the heat exchanger, from which it is carried to the steam generator. (Courtesy Atomic Industrial Forum, Inc.)

and can therefore be used as a nuclear fuel. So as the breeder reactor consumes its own fuel, plutonium, it converts uranium-238 into more plutonium.

Although still at an experimental stage, the breeder reactor has caused a lot of controversy. Some see it as the answer to all our energy needs. Others, pointing to the liquid sodium cooling mechanism and the high long-term toxicity of plutonium, see it as a risky and long-term menace. Despite these fears, breeder research continues. France, with the most aggressive nuclear program in the world, goes forward to make Super-Phénix, the world's first commercial breeder reactor.

The Problems of Nuclear Energy

When the first nuclear reactors became operational in the late 1950s, they were promoted as the miracle power source of the future, delivering an abundance of clean and cheap energy. A nuclear reactor was then a symbol of progress, prosperity, and national pride. By the early 1970s, the lack of public education and sensational reports on potential atomic accidents caused the public to become skeptical and then develop a mistrust and fear of the nuclear energy program. People today are particularly alarmed by three fundamental problems that may very well determine the future of nuclear energy. These are nuclear waste, safety in reactor operation, and the economic cost of nuclear energy.

Nuclear waste, primarily the spent nuclear fuel, is small in volume but contains such toxic elements as strontium, cesium, and plutonium, which remain deadly for thousands of years. A typical nuclear power plant produces only a few hundred pounds of plutonium each year, yet only one grain of this substance if inhaled can cause lung cancer. Since no chemical process can make nuclear waste harmless, it must be isolated safely from any leakage to our environment and secured from any theft by terrorists. At the moment, bundles of spent fuel are submerged in specially designed water pools at the reactor site, the water serving as both coolant and shield. But these pools are becoming overfilled. A more permanent solution must be implemented soon.

Before its permanent disposal, nuclear waste may be put through reprocessing plants. Useful extracts can be recycled as nuclear fuel or as radioisotopes in industry, research, and medical diagnosis and treatment. Nonreusable nuclear waste will then be mixed with a special glass that is unbreakable and cannot dissolve in water—necessary to prevent leaching of nuclear waste. The glass pellets will then be packed in specially designed durable and corrosion-resistant canisters that will be buried in selected underground formations as much as half a mile below the surface. Salt dome deposits and deep granite formations are prospective candidates as nuclear waste "cemeteries." Both are free of water and have existed undisturbed for millions of years, essential qualifications for prospective burial sites.

A solution to the problem of nuclear waste disposal is critical to the future of nuclear energy. But a decision would have to be made even if there were no nuclear reactors. Nuclear waste is

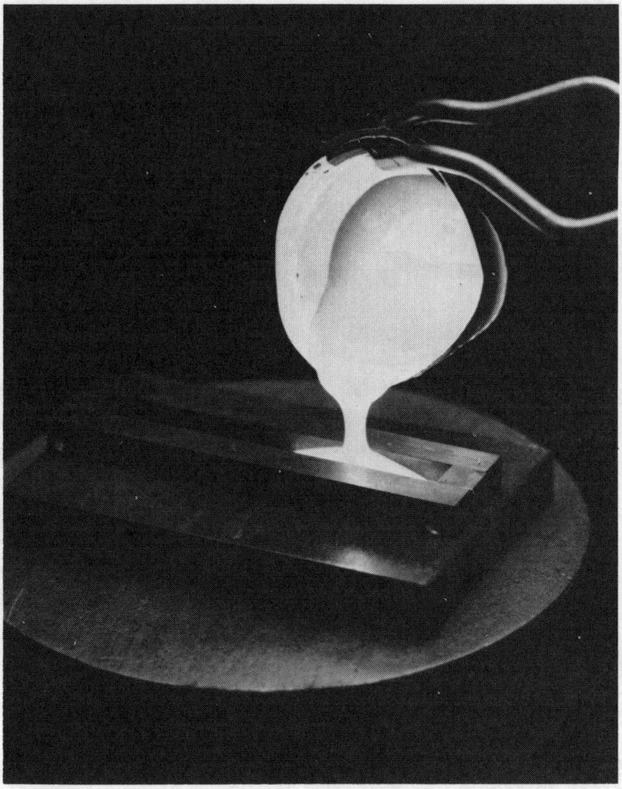

When nuclear waste is mixed with glass it has less chance of leaching into the environment. (Courtesy Atomic Industrial Forum, Inc.)

also produced by nuclear submarines and aircraft carriers, by the nuclear weapons industry, by research institutions, by industry, and by nuclear medicine. Although the technology of nuclear waste management is available, it needs government approval and support, something that politically will continue to drag on for some time. Many politicians, worrying about the long-term ramifications of such a decision, appear reluctant to commit their support to a system that has yet to withstand the test of time. Others fear that reprocessing plants may one day be the target of "nuclear terrorists," who need only steal twenty pounds of plutonium to make an atom bomb. As for the burial site, how does one deal with the social impact of such a decision to the local community, the county, and the state?

An artist's impression of a nuclear waste burial ground. A metal canister containing nuclear waste is lowered through a vertical hole into the tunnel, carved out of solid granite. When the nuclear "cemetery" is filled it will finally be backfilled with crushed rocks and sealed permanently. (Courtesy Atomic Industrial Forum, Inc.)

A possible "cemetery" for nuclear waste. These underground caverns are cut into deposits of basalt, a dense volcanic rock deposited by lava flows millions of years ago. (Courtesy Atomic Industrial Forum, Inc.)

On the morning of March 28, 1979, only two weeks after the release of *The China Syndrome*, a film critical of the operational safety of nuclear reactors, several water pumps at the Three Mile Island (TMI) nuclear reactor in Pennsylvania failed. Through a further combination of human errors and mechanical failures, the temperature of the core increased dangerously to a point that some of its unexposed fuel started to melt. The accident was finally contained without any injuries or major leakages of radioactivity. The shutdown and cleanup costs added up to more than a billion dollars. But the TMI accident caused much more than just a monetary loss. It reinforced a long-held suspicion and public fear that nuclear reactors are dangerous. The fact that not even the experts knew what to do during the crisis damaged the credibility of nuclear personnel training and the Nuclear Regulatory Commission (NRC).

The thorough examination of the whole nuclear energy program that followed the TMI accident revealed flaws in the construction, operation, and administration of nuclear plants. Inspection of reactor records showed that minor breakdowns and leakages had gone unreported, often for fear of a temporary shutdown and loss of revenue. The NRC responded by tightening regulations and inspections, increasing fines, and demanding the installation of new safety equipment at all nuclear reactors, costing their utility company owners billions of dollars. A tougher and longer licensing process, often matched by delays from environmentalists, has almost doubled the completion time of a reactor, often stretching up to fourteen years.

Although no new plants were ordered after the TMI accident, those under construction and already struggling financially felt a devastating economic blow. Escalating inflation and additional red tape that delayed construction skyrocketed their costs from original estimates of a few hundred million to several billion dollars. Desperate utility companies were forced to abandon half-built reactors, losing billions of dollars in the process. These were bad times for nuclear energy. To many, the mighty atom was a god that had fallen from grace, and uncompleted nuclear reactors became, as the *Wall Street Journal* put it, "atomic lemons."

In addition to these problems, common to most nuclear reactors everywhere, the nuclear industry in the United States suffers from two additional weaknesses. Unlike Canada and France, where reactors are standardized, reactors in the United States are custom-built, each different from the others. Lessons learned from the malfunctions of one cannot necessarily be applied to the others. Having to deal with a nuclear reactor on an individual basis makes inspection of operational safety a lengthy and frustrating business. The second problem is caused by the weak connection between the American government and the nuclear industry. Reactors here are owned and operated for profit by utility companies. Nuclear energy programs like those in Canada and France have much stronger government involvement and are administered more efficiently.

But not everything is bad news. There are many examples of the continuous success of nuclear energy. Reactors that for years have been producing low-cost electricity with an untarnished safety record, the impeccable performance of nuclear-powered navy vessels, and the sturdy growth of nuclear reactors in other countries all describe a much rosier picture. France, Taiwan, Finland, Sweden, and Belgium, for example, produce more than 40 percent of their electricity from the atom, as compared to a mere 12 percent in the United States. France plans to raise production to 70 percent by 1990. Limited reserves of fossil fuel, the fear of another oil embargo, and the unpredictable price of oil has convinced many countries that despite the problems, nuclear energy offers at least a partial alternative. The legitimate fears that some third world countries may use nuclear material to manufacture atomic weapons underline the urgent need to strengthen the powers of the International Atomic Energy Agency (IAEA) of the United Nations.

On the morning of April 26, 1986, the nuclear industry's unthinkable nightmare happened. The nuclear reactor at Chernobyl in the Soviet Union went completely out of control. A malfunction in the cooling system caused the overheating of the nuclear fuel and the production of hydrogen gas, which in turn caused a violent explosion. The roof of the reactor building was blown off and radioactive material escaped into the atmosphere. Winds spread this radioactivity over many European countries, which measured unusually large readings of radiation for days. The core of the reactor melted and its graphite pile, used as a moderator in Soviet reactors, was on fire for days. Although only a few people were killed, hundreds were hospitalized for severe radiation sickness, which can be fatal. Thousands were evacuated and some may never be allowed back since part of the area may be sealed off for years. Thousands are expected to develop radioactivity-induced cancer in the next twenty years.

It is unfortunate that the Chernobyl reactor, like most Soviet reactors, was not housed in a concrete-steel dome, standard in nuclear reactors throughout the world. Such a dome can withstand a chemical explosion and can, therefore, prevent the spewing of radioactivity into the atmosphere. Apart from causing human suffering and an enormous financial setback (it will cost the Soviets billions of dollars in clean-up and in reactor modifications), the Chernobyl incident stands as a painful reminder of the realities and dangers of the nuclear industry as well as of the urgent need to take action on certain key issues. The drifting radioactive cloud reminded all of us that reactor safety is no longer a local or a national concern but a global one. Therefore an international body of scientists and engineers is needed to monitor safety in nuclear reactors everywhere. Better communications among nuclear nations, the establishment of universal guidelines, and more coordinated efforts, particularly in emergency cases, are essential. In an atmosphere of international cooperation, many lingering problems such as those of the reprocessing and disposal of nuclear waste may finally be solved.

In addition to the work being done on correcting the many problems of nuclear fission, another form of nuclear energy is currently enjoying tremendous research interest and financial support: nuclear fusion. Although it is still in its technological infancy, and with monumental research obstacles, the many advantages of this new type of energy make it an irresistible venture.

Nuclear Fusion

Nuclear fusion is a high-temperature process in which two light nuclei are fused to form a new nucleus. During the process some matter is converted into energy. Unlike fission, fusion nuclear fuel is in great abundance; it is very cheap and is nonradioactive. It is made of the hydrogen isotopes of protium, deuterium, and tritium—hydrogen atoms of slightly different mass because of an extra neutron in their nucleus. Since water is primarily made of hydrogen, vast amounts of protium and deuterium can be derived from the oceans. Although there is only one deuterium atom in every 6,600 atoms of ordinary hydrogen, the oceans can provide fusion fuel for millions of years. Tritium, on the other hand, is produced from lithium, an element with deposits sufficiently large to provide at least a 100,000-year supply.

A major attraction of fusion is the mild radioactivity of its waste products. Unlike fission nuclear waste, which requires safe storage for thousands of years, fusion nuclear waste is smaller in quantity and requires storage for only twenty to thirty years, making its disposal easier, safer, and more economical.

Although billions of dollars have been spent so far in fusion research, its future is by no means certain. The technological obstacles remain as enormous today as they were twenty years ago. But they are now attacked with more sophisticated tools. The skeptics fear that the technological difficulties will prevent fusion from ever becoming economically viable or even technologically possible. The optimists, on the other hand, are convinced that commercial fusion energy will be available by the middle of the next century. We know that fusion can work. After all, it is the source of power for the stars. Artificial fusion has been achieved on earth, but only in the momentary and uncontrolled explosion of the hydrogen bomb.

To achieve fusion, the nuclei must first be brought together. But when nuclei come close together, they repel one another because of their similar electric charge. Heating them increases their speed and so brings them closer together—but the closer they get, the more powerful the repulsive force gets. To overcome repulsion, an even higher temperature is required. To bring the nuclei close enough to fuse, temperatures of 60 to 100 million degrees Centigrade are needed. At such temperatures the fuel, now a superhot and electrically charged gas called plasma, can instantly vaporize a container of any material with which it comes in contact. If a strong and appropriately shaped magnetic field is used, however,

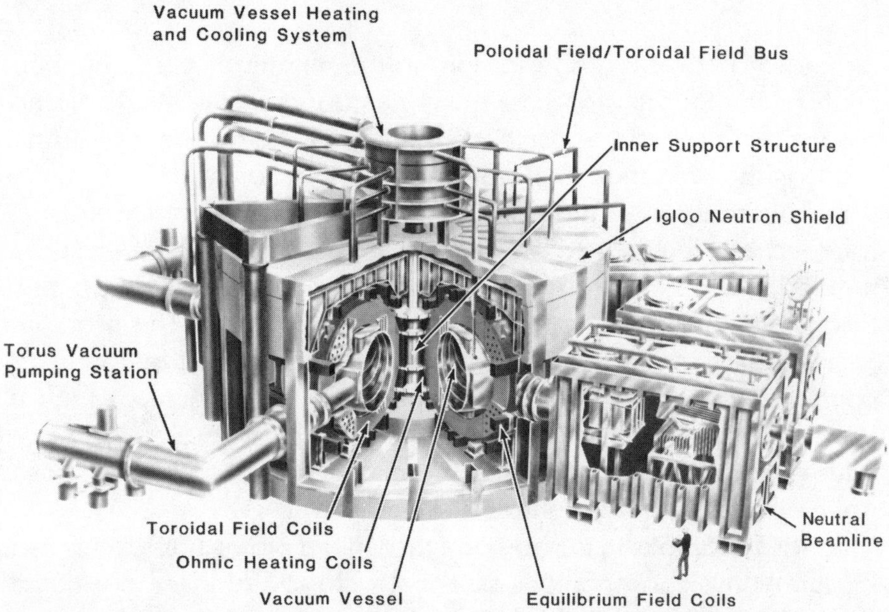

The Tokamak Fusion Test Reactor at the Princeton Plasma Physics Laboratory. (Courtesy Princeton Plasma Physics Laboratory)

Workers inside the vacuum vessel of the Tokamak Fusion Test Reactor at Princeton. (Courtesy Princeton Plasma Physics Laboratory)

the plasma can be made to float in the middle of the container. Such a configuration is known as a magnetic bottle. Of all the many different designs that have been tried, the most popular one today is the Tokamak, first developed in the Soviet Union.

A Tokamak is a doughnut-shaped evacuated cylinder surrounded by electrical coils that produce the strong magnetic field needed to suspend the plasma. In a working system the magnetic field is also used to compress and heat the plasma. The problem is that the hotter the plasma gets, the more difficult it becomes to control it magnetically, which often leads to a breakdown. The use of superconducting coils carrying much more electricity because of their supercold temperatures has improved plasma control by boosting the magnetic field.

So far Tokamaks in Europe and the United States have achieved the three physical conditions essential for fusion: high temperatures (from 60 to 100 million degrees Centigrade), plasma density (3×10^{14} particles per cubic centimeter), and duration (a minimum of one second). Unfortunately, these have not been all achieved simultaneously, as required for a complete process. Another problem, of an economic nature, is achieving a break-even situation—one in which the energy released by fusion is at least equal to the energy expended. Recent advances at the Tokamak at Princeton indicate that a break-even situation may be reached within the next ten years.

The use of lasers as an alternative method to Tokamak has been developed during the last few years. In this method, pinhead-size glass microballoons filled with solid fusion fuel are crashed by powerful light pulses from giant lasers. The crashing pressure developed by these pulses is enormous—on the order of 10^{12} (a million million) times atmospheric pressure. The implosion, which lasts less than a billionth of a second, raises the temperature of the fuel to 100 million degrees Centigrade and its density to 10^{26} atoms per cubic centimeter. Because the duration of the implosion is so short, the inertia of the fuel holds it in one place during the process. The method is hence called inertial confinement. Compared to the Tokamak, where fusion is a continuous process, inertial confinement resembles the firing of a series of miniature hydrogen bombs.

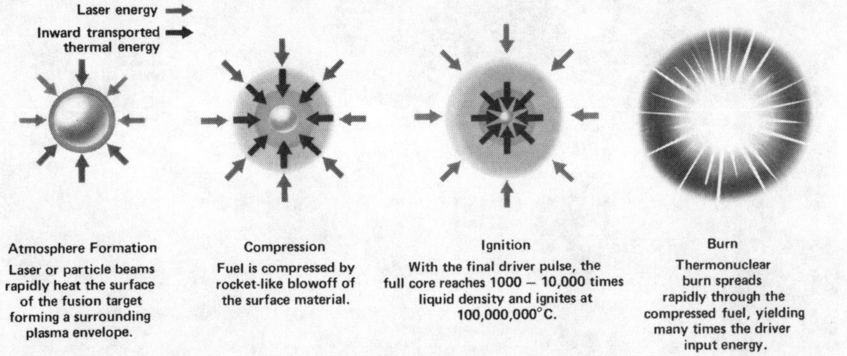

The stages of nuclear fusion by the laser inertial confinement method. (Courtesy Lawrence Livermore National Laboratory)

Until 1980, Shiva was the largest laser fusion project at the Lawrence Livermore National Laboratory. Named after the Hindu god with four arms, Shiva had twenty neodymium-glass lasers delivering a power of more than 30 trillion watts in less than a billionth of a second. Other laser fusion projects, such as Antares and Nova, currently the largest project, are of a different design and use more powerful lasers made of carbon dioxide gas. In some other systems, lasers are replaced by particle beams. Although all of these are very promising projects, they are still at the research stage. For example, Shiva was far away from the break-even situation. Its peak efficiency was less than 2 percent, which means that for every one hundred units of energy invested only two were returned. In a commercially successful laser fusion reactor, the system must be fired ten times per second. Shiva could at best be fired only once every four hours.

Planning for tomorrow's energy is both politically and economically a tough decision that needs to be made soon. But what criteria must we use to make such a decision? Low cost, availability, and a clean environment are the dominating factors. Antinuclear groups point to the accident at Chernobyl, the long-term dangers of nuclear waste, and the escalating costs of nuclear reactors. Environmentalists, on the other hand, point to the acid rain that is produced by sulfur dioxide released in the burning of coal. Solar energy, clean and inexhaustible, is becoming less and

Shiva, a giant laser system used for fusion research at Lawrence Livermore Laboratory. It was replaced by Nova, an even more powerful system. (Courtesy Lawrence Livermore National Laboratory)

less expensive and offers a very realistic alternative. But its limitations are obvious. Without the sun, at night and on cloudy days, its production stops. Its best future use, therefore, is to provide not all but a considerable portion of our energy needs. The same may be true for some other exciting new sources, such as geothermal, wind, and tide energy, currently under research and development.

It appears that our best strategy for tomorrow's energy is a three-part package that includes nuclear energy. The first part of our future policy is to diversify our energy sources as much as possible to avoid the risk of having all our eggs in one basket. The oil embargo of the 1970s is a strong reminder of such vulnerability. Nuclear energy is one such basket, in which we already have placed some of our eggs. The second part is to improve energy conservation and reduce energy-related pollution. Finally, we must make a commitment to an aggressive research program on new sources of energy, particularly solar power and nuclear fusion.

The unfortunate part of the energy story is that although the

Nova, the world's most powerful laser. A technician makes adjustments on a laser amplifier. Nova can generate laser beams powerful enough to implode a fuel pellet (a tiny hollow sphere containing deuterium and tritium) and create a miniature star—a fusion reaction. (Courtesy Lawrence Livermore National Laboratory)

Laser technicians install diagnostic instruments on the Nova target chamber, a 16-foot-diameter aluminum sphere inside of which the ten powerful laser beams simultaneously converge on a small fuel pellet to produce fusion reactions. (Courtesy Lawrence Livermore National Laboratory)

A technician adjusts one of dozens of "box" laser amplifiers used in the ten arms of the Nova laser. (Courtesy Lawrence Livermore National Laboratory)

crisis is a global one, industrial nations have failed miserably in forming an alliance to tackle the problem on a united front. Many countries could have shared the burden of research costs and could have committed their best scientists and engineers to such an international task force. A global cooperative program of this kind could do much more than provide us with energy in the future. It could help break down some of the political and economic fences that nations have built around themselves.

Chapter Four

The Silicon Chip

It is so tiny that it cannot be seen even with the most powerful microscope, yet it is the giant on which all our modern technology is founded. There are zillions of them in a piece of wire. When they are properly "motivated" to run their erratic race, they create conditions that give rise to hundreds of crucial functions, critical to our modern way of life. Squeezed through thin wires, they generate heat and light. Forced through coils inside a magnet, they generate motion. Properly tapped, they can carry messages, speech, and music. They are responsible for the communication between brain and muscles, and they even transmit experiences to the brain's memory. Like a painter's brush, they put together pictures inside a box. There is hardly a modern appliance that does not depend on their peculiar properties. Computers cannot be made or function without them. They are indeed our best friend: the electrons.

Although they were identified as individual particles in 1896, they were put into use long before that. When electricity was developed in the nineteenth century, it soon became the major source of energy, pushing industrial production to new heights. Its multitude of uses has literally transformed the landscape. Poles were erected to carry it, power plants were built to produce it, and appliances were built to use it. Experimental and theoretical research in electricity gave rise to many discoveries that started fields of their own. X-rays, radio waves, electronics, and solid-state physics (the mother science of semiconductors, which make up computers) were a few of these.

But the most dramatic impact of electricity is found in the field of information technology. From Morse's telegraph to weather satellites, from Bell's telephone to radio and television and now the computer, these "children" of electricity have reshaped the human landscape more than any other technology. They brought the information age, which can elevate us to higher levels of civilized existence. Robots, computers, and satellites, technologies that were unthinkable a century ago, are forging a new human species. Besides their obvious technological benefits, they are bringing a new freedom from boring and tedious tasks, creating the environment in which generations to come will be able to develop their minds more and more. On the optimistic side, they may even bring about a global identity for all—one that could eradicate many of the hostilities and injustices that exist today.

In the center of this new age lies the computer. Its very beginnings go back to the early work on electricity. Out of this field came the many tools that brought about the information age.

Early Milestones in the Development of Electricity and Telecommunications

To the ancient Greeks, static electricity and magnetism were nothing but fascinating phenomena. Thales of Miletus, one of the "seven wise men" of ancient Greece, knew that amber, when rubbed, attracts such light bodies as hair, chaff, and straws. The Greek word for amber is *elektron,* out of which the modern terms electron and electricity are derived.

Electricity and magnetism remained fascinating mysteries of nature for hundred of years. The first scientific explanation for magnetism was offered by Dr. William Gilbert (1544–1603) in his classic book *De Magnete*, published in 1600. Dr. Gilbert, a physician to Queen Elizabeth I, explained many magnetic phenomena and proposed that the earth was a giant magnet. His interest and research activities extended to electricity, and he often presented fascinating demonstrations of electricity and magnetism to the English royal court.

In the eighteenth century simple hand-driven machines were built that could produce electricity by rubbing cloth or fur with glass or a resin-type material. Often the electricity produced was stored in a special jar—the Leyden jar—that was made of glass lined with metal foil. Leyden jars were made to discharge instantly if they were touched by another metal, and the discharge was accompanied by sparks. At about the same time—almost two hundred years after Gilbert—Benjamin Franklin (1706–90) was making some interesting discoveries. In his famous experiment with a kite he showed that lightning is a form of electricity. He proposed that the erection of lightning rods on tall buildings and on the masts of ships could draw the electricity from the clouds harmlessly into the ground or water. Interestingly, some of the clergy opposed the use of the lightning rod, arguing that lightning expressed the wrath of God and that lightning conducted into the earth via a lightning rod could trigger earthquakes. Franklin also proposed a theory of electricity in which all electrically neutral bodies could develop an electrical charge by friction. In some cases, by his explanation, friction would add electricity to objects and make them "positive"; in others, it could remove electricity and make them "negative."

The last quarter of the eighteenth century witnessed a vigorous scientific controversy over the topic of "animal electricity." The question of how electricity is produced in animal tissue was to lead later to the construction of the first battery. Luigi Galvani (1737–98), a professor of anatomy at the University of Bologna, noticed that dissected frogs would flex their muscles if they were touched by a variety of metals. His theory that the legs contained a fluid he called "animal electricity" was later challenged by his Italian compatriot Volta.

Alessandro Volta (1745–1827), a professor of physics at Pavia, proposed instead that the electricity causing the legs of the frog to flex was produced by connecting two dissimilar metals through the saline solution contained in the legs. Volta experimentally verified that when two different metals are immersed in an acetic solution they can produce strong electric currents. In 1800 he constructed the first battery, made of a pile of copper and zinc disks separated by paper moistened with brine or acid solution.

The invention of the battery marks a turning point in the history of electricity. When static electricity, the kind produced by friction, begins to flow it does so instantly, producing sparks, like a miniature lightning bolt. This prevents the study of the flow of electricity. On the other hand, electricity produced by chemical reactions, as in batteries, can be made to flow at a controlled rate. A steady and continuous flow enabled scientists to study the properties of electricity in a variety of experiments. Thanks to these experiments, many electrical machines and instruments were developed. The two that stand above all the others in these early pages of the history of electricity are the electric motor and the dynamo, or generator.

Hans Christian Ørstead (1777–1851), a Danish physicist and professor at Copenhagen, was a disciple of the German philosophical school that proposed a connection between light, electricity, magnetism, and chemical forces. Ørstead searched for the connection between electricity and magnetism, and in 1820 he found it. He discovered that electricity flowing through a wire generates a magnetic field around it, "radiating" out of the wire in strengths of circular symmetry. This newly discovered property of electricity gave the Englishman Michael Faraday a valuable clue in his attempts to harness electricity.

Historians of science call Faraday the greatest of all experimental physicists. When as a young man he was hired by the prestigious Royal Institute in London, it was to wash bottles and clean up the laboratories. Although his education was only of elementary-school level, he was driven by a ferocious thirst for knowledge and an obsession with experimental research in the chemical and electrical sciences. His scientific accomplishments were many. He discovered benzene, and he was the first to liquefy gases. He discovered the laws of electrolysis, he made stainless steel, and he explained many of the properties of polarized light. But his most noted contribution was the invention of the motor and the dynamo.

Faraday repeated Ørstead's experiments to find out for himself the connection between electricity and magnetism. He extended his work further because he felt that the magnetism produced by an electrical wire could be put to some practical use. He discovered that a wire suspended between the poles of a horseshoe magnet experienced a force every time electricity flowed through it. The magnitude of the force depended on the amount of the electric current and the strength of the magnet, and it could be powerful enough to kick the wire out of the magnet. Using an ingenious device made of a wire floating in mercury, Faraday was able to convert this linear and sudden force into a smooth and rotational motion. So the electric motor was born in 1821.

Faraday was fascinated by the simplicity of the principle behind the motor. The heart of the principle was Ørstead's discovery that the flow of electricity produces magnetism. In the motor, the motion is due to the interaction of two magnetic fields—that of the magnet and of the electric coil. But Faraday went on to the next obvious question: If electricity can produce magnetism, can the reverse be true? Can magnetism generate electricity? In October 1831, after ten years of sporadic trials, he finally succeeded. The phenomenon, known as electromagnetic induction, is extremely simple. When a magnet moves in and out of a coil, electricity is generated and flows through the coil. Since linear motion is awkward, the next improvement was one of design: the motion converted to rotational. This new invention was the dynamo, or electric generator, destined to dominate the power industry.

Like many new inventions, the motor and the dynamo were not received with instant enthusiasm. Engineers, industrialists, scientists, and politicians saw these new inventions as mere scientific curiosities or toys, without any prospect of practical use. It is said that when England's minister of finance was visiting Faraday's laboratory, which was funded by the state, he asked what possible use such devices could have. Faraday, disappointed by his shortsightedness, gave an explanation the minister could understand—

that one day the state might raise considerable revenue by taxing these devices.

Michael Faraday, the "prince of experimenters," is regarded also as the father of a theoretical concept of physics that makes him a bold precursor of Einstein. His ignorance of mathematics forced him to create tangible explanations as to how magnetism is propagated through space. He pictured "lines of force" extending from a magnet throughout space in the form of a field. He further proposed that all space was filled with magnetic, electric, radiant, thermal, and gravitational forces. It was out of such ideas that Maxwell formulated the famous electromagnetic theory that describes electricity and magnetism mathematically. Maxwell also deduced that light is electromagnetic radiation and predicted the existence of invisible electromagnetic waves. The concept of the field was a keystone in Einstein's general theory of relativity.

The significance of Faraday's work is paramount. Yet this giant scientist remained humble all his life. A blacksmith's son, he lived simply and died in poverty. He could have made a fortune from his inventions, but he intentionally dissociated himself from the commercial prospects of his discoveries. His disdain of wealth was only a part of his admirable scientific ethical code. He shared his knowledge of science by giving popular lectures with lots of demonstrations to the public, particularly schoolchildren. Concerned about the horrors of environmental pollution, so characteristic of Victorian London, he expressed the need for a cleanup in a letter to the *Times* of London. But above all he refused to associate himself or his science with the military machine. He rejected his government's pleas to help manufacture poisonous gases for military use. It is unfortunate that although Faraday is acknowledged for his great scientific contributions, his marvelous human qualities are so often neglected or completely ignored.

Faraday started the electrical revolution that quickly penetrated all aspects of everyday life. Electricity was first used to provide power for industry in place of steam. As years went by its use extended to such other areas as transportation, communications, medicine, and household appliances. The world of electricity brought excitement to all. To some, this was also a great opportunity to tap an unexplored and versatile new technology. There were electrical inventions everywhere. Some were bizarre and totally useless, others simple and practical. Still others were critical in the further development of a variety of new technologies.

It is interesting that the fathers of some of these early electrical inventions were not "ivory tower" university professors but rather uneducated wizards from all ways of life. Their curiosity and ingenuity, often fueled by a desire for financial success, produced such inventions as the telegraph, the telephone, the light bulb, and the radio. Thomas Alva Edison (1847–1931)—often called "America's most useful citizen"—had more than twelve hundred patents in his name. He changed the world when on September 4, 1882, he introduced the first large-scale electric lighting system. Among his great inventions were the light bulb, the phonograph, the movie camera, and the rechargeable alkaline battery.

Some of these early electrical inventions played a key role in the field of communications. These gadgets, primitive as they may appear by modern standards, were critical in the establishment of the speedy and mass communications system so essential in the information age.

Until the late nineteenth century long-distance communication was a slow, difficult, and expensive task that required messengers on foot, horse, carriage, or ship. The flow of electricity along a wire inspired many scientists at that time to use it to transmit messages. A variety of exotic telegraph systems were tried, but all were unreliable and impractical. In one such system, twenty-six pith balls were suspended in contact with as many wires, each representing a letter of the alphabet. When electricity flowed through one wire it set its corresponding ball vibrating by electrical discharge. In another system, the wires ended in tubes containing a clear solution. Every time a letter was transmitted a tube produced bubbles by electrolysis. In yet another case, the transmitted electricity made magnetic needles turn and point to the letters of the alphabet. Although all of these devices were based on good and simple science, they were nevertheless examples of poor technology that was either impractical or unprofitable.

It was Samuel F. B. Morse (1791–1872), a painter, who produced both the first mechanism and the first convenient code that made telegraphy a practical reality. In the Morse code, each letter of the alphabet is represented by a combination of dots and dashes, produced by short and long pulses of electricity. Morse is an example of a paradox that was so characteristic of those times. He was neither a scientist nor a technologist. Not only was Morse ignorant about electricity and magnetism, he was also an incompetent mechanic and a poor businessman. Yet despite his background he was fascinated by the idea of using electricity to transmit messages. He survived many personal and financial difficulties while seeking recognition and financial support for his invention. After much debate, the U.S. Congress finally approved a small grant for a small-scale trial of Morse's system of telegraphy. Soon after that everyone realized that the new invention was an invaluable tool.

But messages sent by dots and dashes were slow and also lacked the personalized flavor of voice communications. Alexander Graham Bell, a doctor, initially used his expertise in speech, acoustics, and the human eardrum trying to bring communication skills to the deaf and dumb. His major technological task was to create a "sound-shaped" electric current. He found some of the answers to his technical problems in the works and advice of Joseph Henry (1797–1878), a brilliant American physicist with many electrical innovations to his credit.

Bell's telephone had two basic parts: a mouthpiece and an earpiece. Words spoken at a metal diaphragm in the mouthpiece caused disturbances in the tiny electric current that flowed from one telephone to another. On their arrival at the earpiece on the other end of the line, these electrical disturbances set another diaphragm vibrating with the help of an electromagnet. These vibrations reproduced the original message.

But Bell did not limit himself to the use of electricity as the only carrier of sound. In 1880, only four years after he invented the telephone, he received a patent for the photophone, a remarkable device in which rays of light replaced electrical wires in the transmission of messages. The idea was simple. A beam of sunlight was made to bounce from a mirror attached to a mechanism that vibrated in response to speech. This caused the light beam to vary in intensity. At the receiving end, a detector translated these var-

Alexander Graham Bell tests the harmonic telegraph. In June 1875, Bell heard a reed's twang transmitted over a wire; it confirmed his transmission theory. (Courtesy Bell Laboratories)

iations into sound. Unfortunately, the photophone was an idea a hundred years ahead of its time. The technological difficulties of using light to carry messages were overcome when a new miracle tool was invented in 1960. That was the laser.

Maxwell's electromagnetic theory predicted invisible electromagnetic waves. The German physicist Heinrich Hertz demonstrated in 1886 that such waves could be produced and detected. His experiment was crude and simple but very convincing. When two high-voltage plates were made to spark, they generated electromagnetic waves that were transmitted in all directions through space. They were detected by a wire loop with its ends open and

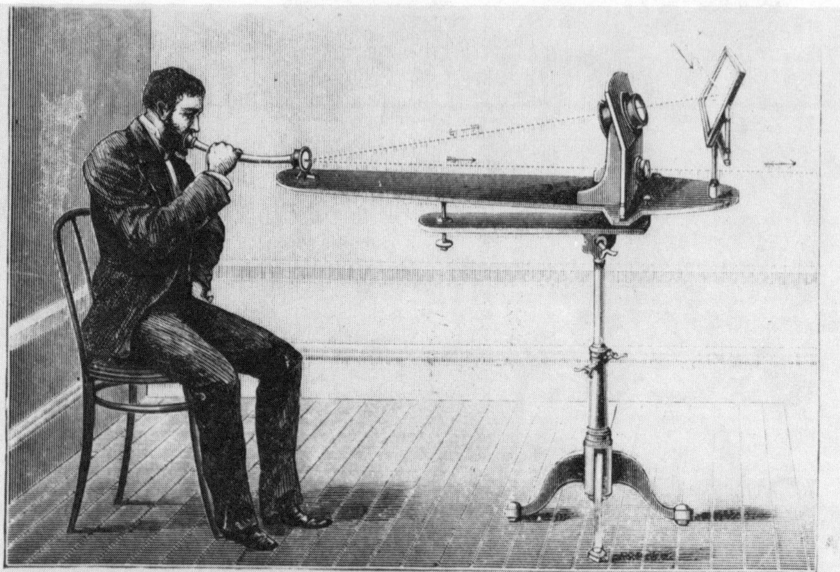

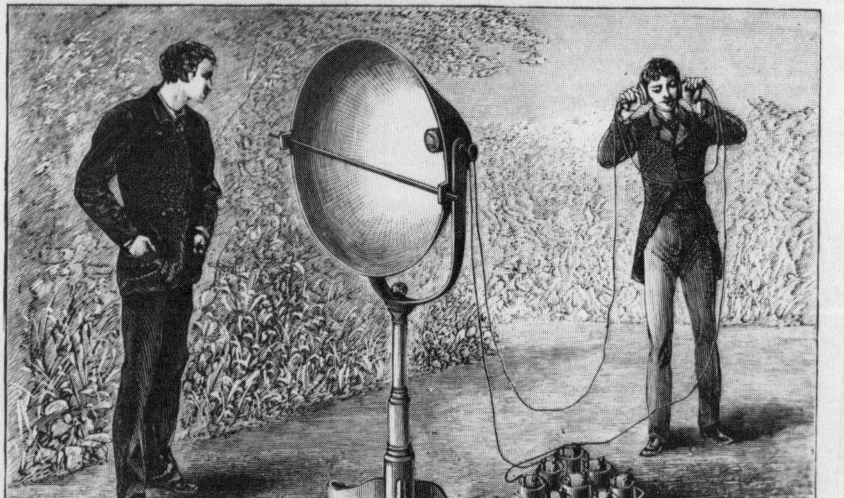

In Bell's "photophone," sunlight was bounced from a reflector through a lens to a mechanism that vibrated in response to speech *(top).* This caused the light beam to vary in intensity. At the receiving end, a selenium detector translated these variations into electrical current to recreate speech through a telephone receiver *(bottom).* (Courtesy Bell Laboratories)

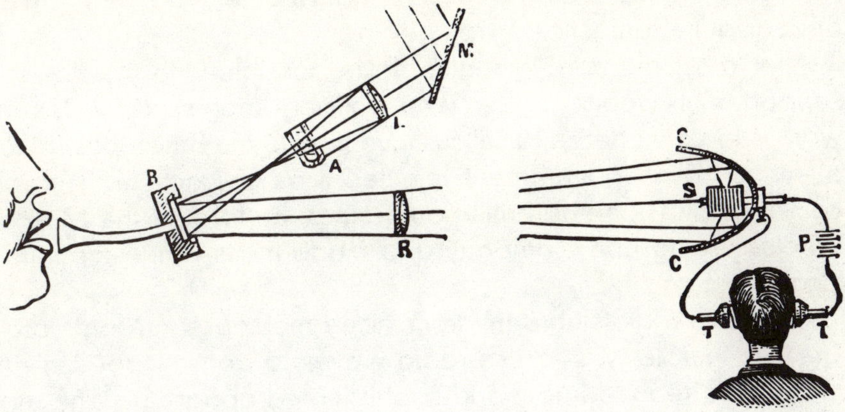

The apparatus of Bell's photophone, when used for talking with reflected sunlight. (Courtesy Bell Laboratories)

slightly apart. When electromagnetic waves crossed through the loop, they produced tiny sparks across the gap. Other physicists followed up, extending Hertz's work. More elaborate systems to produce and detect electromagnetic waves were made, and new theories were proposed. But it was the young Italian Marconi, a layman rather than a scientist, who put all these new discoveries into a new revolutionary technology: radio communications.

Guglielmo Marconi (1847–1937) never studied at a university, and neither did he show any special interest in science or technology. He learned about electromagnetic waves through his friendship with Professor Richi, a physicist and a neighbor of the Marconis at their summer home in Bologna. The young Marconi was captivated by these invisible waves. He soon visualized their great potential in communications.

"It seemed to me," Marconi said in a lecture years later, "that if the radiation could be increased, developed, and controlled it would be possible to signal across space for considerable distances. My chief trouble was that the idea was so elementary, so simple in logic, that it seemed difficult to believe no one else had thought of putting it into practice. I argued there must be more mature scientists who had followed the same line of thought and arrived at almost similar conclusions. From the first the idea was so real to me that I did not realize that to others the theory might appear quite fantastic."[10]

After a lot of improvements, made possible with the financial support of his wealthy father, who was very impressed by his son's work, Marconi offered to sell his invention to the British post office, after the Italian government rejected it as impractical. With encouragement from the British authorities, he formed the Marconi Wireless Telegraph Company and made mass wireless communications a reality.

Marconi's radiotelegraphy transmitted messages in Morse code. The next hurdle was to use radio waves to transmit speech and music. This required new and sophisticated apparatus. The most critical part was the triode valve, a device that amplifies the electromagnetic waves that are weak on arrival. The American physicist Lee De Forest (1873–1961) is recognized as the chief inventor of this new electronic device. With the introduction of the first commercial radio stations in the 1920s and the first commercial television in the 1940s, the initial round of electronic communications was completed.

The telegraph and the telephone had a dramatic impact on lifestyles. Distance and time suddenly shrank in this new world of instant communications. But the effects of both were minute compared to those of radio and television. Never before was there so much free information available at the push of a button and the turn of a dial. Education, entertainment, news, and music were more democratically distributed through the airways and were no longer the privilege of the rich. But the manipulation of information or propaganda, whether for political, commercial, or religious reasons, became more powerful also. People learned to accept the mixed blessings of all technologies. A conscious effort is made to at least be aware of the dual role of technology in our lives. We must be even more watchful of this now with the emergence of an even more powerful information technology. Based on the coming together of the television, satellite, and computer, a second round of information technology has begun. These three powerful new instruments have improved tremendously through the magic of the transistorized silicon chip, its function based on the unique flow of electricity through materials known as semiconductors.

The Transistor

On December 23, 1947, after almost two years of work, a dramatic discovery was coming to completion in the Bell Telephone Laboratories in New Jersey. Three physicists, John Bardeen (1908–), Walter H. Brattain (1902–), and William Shockley (1910–), who eight years later shared the Nobel Prize in physics for their invention, were to be proud parents of one of the most powerful devices ever invented, the transistor. This tiny device, robust and cheap to make and operate, was destined to begin a new phase in the era of information technology.

Understanding the transistor requires first some basic understanding of semiconductors. These are materials that are neither conductors nor insulators of electricity but something in between. Whether a material is a conductor or an insulator depends on its atomic structure, for it is the arrangement of the electrons in shells around the nucleus that determines the physical properties of materials. The number of electrons in the outer shell of an atom, often referred to as valence electrons, and their energy characteristics determine the chemical properties and electrical conductivity of the material.

The transistor's inventors. *Left to right:* William Shockley, Walter H. Brattain, and John Bardeen, in 1948. (Courtesy Bell Laboratories)

A typical atom used in the manufacture of semiconductors is silicon. It has fourteen electrons that are distributed in three shells around the nucleus. The shell nearest to the nucleus, the K shell, can have only two electrons, whereas each of the other two, the L and M shells, can have up to eight electrons. In silicon, the M shell therefore has only four electrons, so the atom has the tendency either to borrow four more electrons to complete its shell or to lend out its four electrons. When silicon atoms get together to form crystals, they share each other's outer electrons, resulting in a material that has all its electrons strongly bound in shells.

Since silicon crystals have no loosely bound electrons, which are responsible for the conduction of electricity, they behave as insulators. If small quantities of materials with atoms of five or three electrons in their outer shells are added during the crystallization process, silicon becomes a semiconductor. For example, if phosphorous, which has five electrons in its outer shell, is added to silicon as an impurity (a process called doping), the resultant crystal has one loosely bound electron for each phosphorous atom added. This impure silicon is now called an n-type semiconductor, the n signifying an easy availability of negative electric charges. If boron, with three electrons in its outer shell, is added as an impurity, there will be a vacancy, or "hole," for one electron for each boron atom added. This is a p-type semiconductor, where p stands for positive electric charge. Both types of semiconductors

The first transistor ever assembled; the year was 1947. Its inventors shared a Nobel Prize in physics. (Courtesy Bell Laboratories)

can carry electricity. The ease of conduction depends on the amount of the impurity added. Usually the ratio is about one impurity atom to every few million silicon atoms.

But the usefulness of semiconductors is not so much their ability to conduct electricity as their ability to control its flow. The first use of semiconductors was in the manufacturing of diodes, devices that allow electricity to flow in only one direction. This was previously done by vacuum tube diode valves, which not only were bulky, fragile, and expensive, but also consumed a lot of electrical energy. The principle of operation for both types of diodes is that electrons will flow only toward a positive terminal. The following diagram explains this:

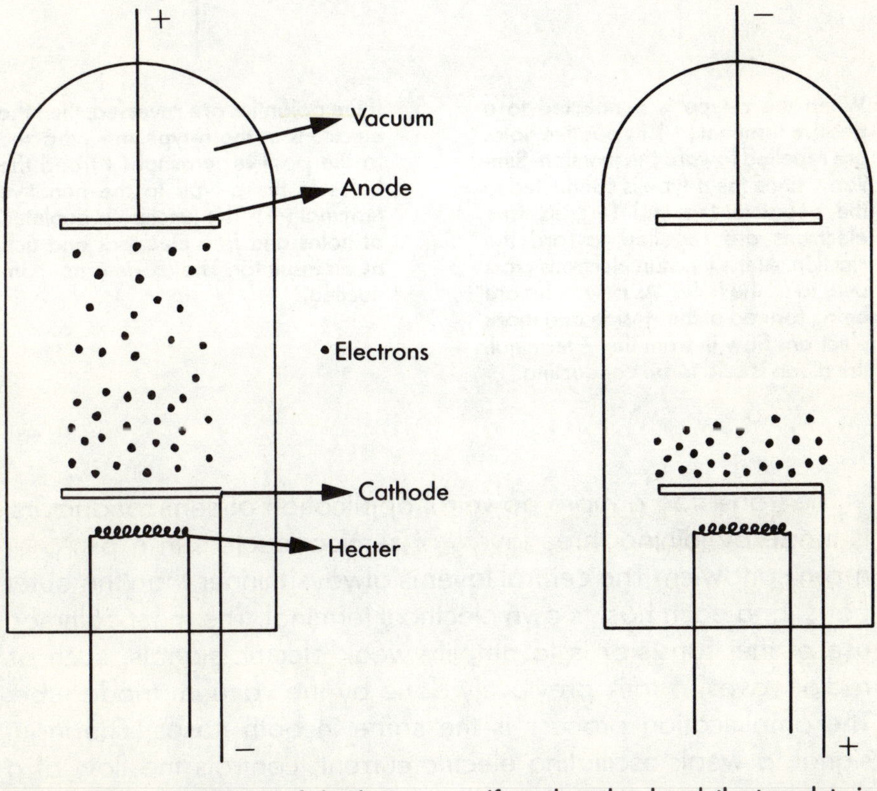

Electrons are "born" out of the hot bottom plate (cathode). If the top plate (anode) is positive, they will flow to it. The diode is said to be conducting.

If, on the other hand, the top plate is negative and the bottom is positive, the electrons stay near the bottom plate. The diode does not conduct.

When a p- and an n-type semiconductor join, they form a pn junction diode:

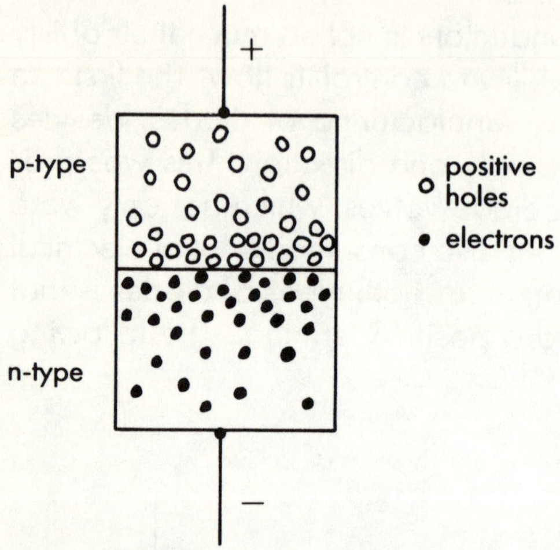

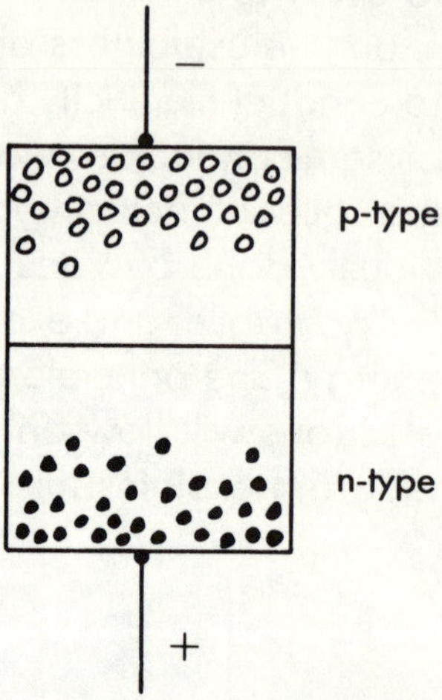

When the p-type is connected to a positive terminal (+), its positive holes are repelled toward the junction. Similarly, since the n-type is connected to the negative terminal (−), its free electrons are repelled toward the junction. At the junction electrons cross over to fill the holes. As new holes are being formed at the + side and more electrons flow in from the − terminal, the diode is said to be conducting.

If the polarities are reversed, then the electrons in the n-type are attracted to the positive terminal (+) and the holes in the p-type to the negative terminal (−). The junction is depleted of holes and free electrons and acts as an insulator. The diode is not conducting.

The transistor, a more powerful application of semiconductors, is made by joining three layers of semiconductors in a p-n-p or n-p-n sandwich. The central layer is always thinner than the outer parts, and each has its own electrical terminal. The most common use of the transistor is to amplify weak electric signals, such as radio waves, a task previously done by the vacuum triode tube. The amplification process is the same in both cases. The input signal, a weak oscillating electric current, controls the flow of a much larger current. Every oscillation in the input signal is repeated in the main current. The outgoing signal is therefore an exact copy of the input signal but at a higher intensity. The following diagram shows how a signal is amplified with a triode valve and an n-p-n transistor:

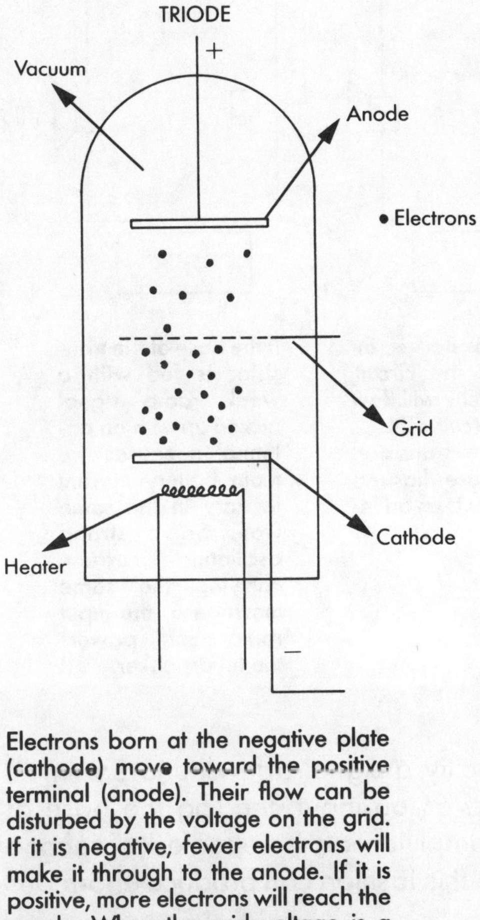

Electrons born at the negative plate (cathode) move toward the positive terminal (anode). Their flow can be disturbed by the voltage on the grid. If it is negative, fewer electrons will make it through to the anode. If it is positive, more electrons will reach the anode. When the grid voltage is a weak oscillating signal, it forces the flow of electrons to oscillate in unison, resulting in an amplified signal.

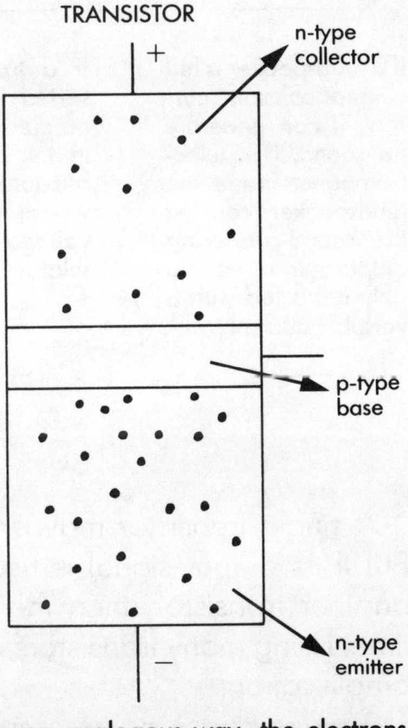

In an analogous way, the electrons flowing from the negative n terminal to the positive n terminal can have their flow interfered by the voltage on the p-type central part, known as the base. If the base voltage is a weak oscillating signal, it forces the large current to oscillate in the same manner, causing an amplification.

A crude example of the amplifying power of the transistor is shown in the following case:

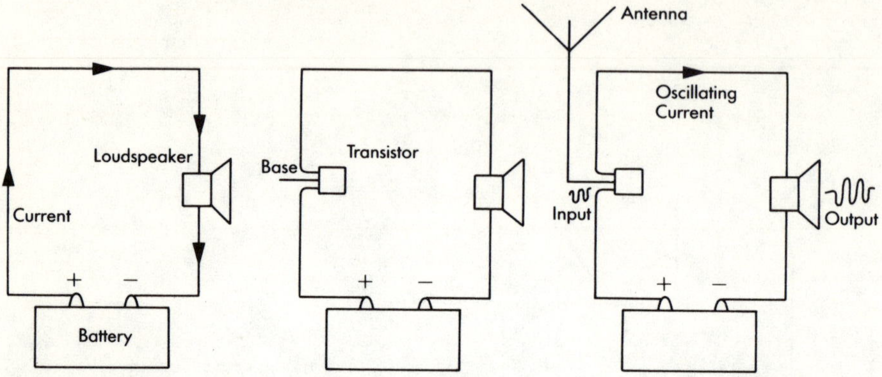

If a loudspeaker is fed with a *constant* current, it can generate no sound. (The electromagnet inside the loudspeaker can set the sound-producing diaphragm to vibrate only if it is fed with a variable current.)

If a transistor is inserted in the circuit, no electricity will flow in the circuit. This is because a transistor, whose base has no voltage, acts as an insulator.

If the base of the transistor is fed with a weak radio signal picked up with an antenna, it causes the main battery current to vary in the same way. So a strong oscillating current, carrying the same message as the input radio signal, powers the loudspeaker.

A single transistor may amplify a signal a few hundred times. But if its output signal is used as an input, feeding the base of another transistor, then the amplification is repeated a second time. Using many transistors in this fashion can produce enormous amplification.

Besides its use as an amplifier, the transistor is also used as a switch. When its base, the thin central layer, has no voltage, the transistor does not allow current to flow through it. In that state the transistor acts as a closed switch. In computers the complex process of storing and retrieving information boils down to the simple principle of having switches turned on and off. Millions of minute transistors, etched on tiny silicon chips, have become the heart of the modern computer.

The discovery of the transistor in 1947 started an electronic revolution. Electrons in a tiny sandwich of semiconductors could outperform the vacuum tube electrons. The transistor could amplify, oscillate, and switch electric signals faster, more efficiently, and more economically than the vacuum tube. Today there is hardly an electronic gadget that does not use transistors. From household appliances to orbiting satellites, from hearing aids to pacemakers and computers, they all contain thousands if not millions of these miracle electronic components. The transistor brought about dramatic social changes. In the same way that Gutenberg's invention in 1455 made the printed world available to all, affordable and portable transistor radios have made information available to everyone at the push of a button. In a characteristic statement, Brattain said, "... even the loneliest nomad on the steppes of Asia can have the news of the world just by twisting a dial. He doesn't have to learn to read. And once the common man has a fair chance to learn what's really going on, he has a chance to control his destiny."[11]

The Silicon Chip

Within a decade after its invention, the transistor was reduced to a speck of semiconductor material encased in a pea-size metal can. It continued to shrink, to a point where hundreds of them were soldered with other electrical components on plastic boards no larger than cigar boxes. But electronic engineers still considered these boards bulky and expensive. Furthermore, components and their soldered wiring were often torn off these boards. The solution to these problems was thought out independently by Jack Kilby (1923–) at Texas Instruments and Robert Noyce (1927–) at Fairchild Semiconductors. The idea was to make the electronic components serve as their own circuit board. The new device, called the integrated circuit or chip, was the heart of a new industry—and it sparked a revolution in information technology.

The manufacture of a chip, the design of which may take up to three months, is a step-by-step process resembling microsurgery in the sterile environment of an operating theater. Silicon, the raw material for all chips, is after oxygen the most abundant element on earth and is found in large quantities in quartz rocks and common beach sand. Pure silicon is refined from quartz, melted down, and grown into long crystals, which are sliced into thin, round wafers. What follows is basically a series of processes similar to silk-screening. Masking, spraying, baking—each plays a role in building a complex multilayer electronic circuit on each chip.

The silicon wafers (each having the diameter of a grapefruit) are first sprayed with a film of oxide, which becomes the insulating

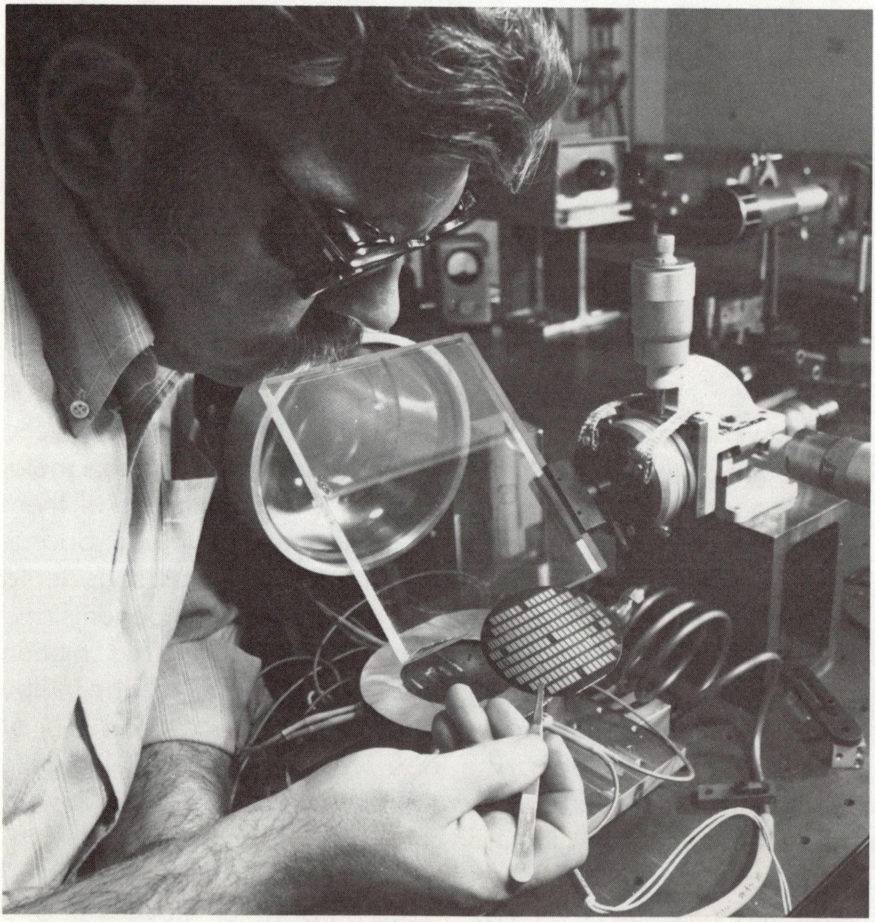

A silicon wafer ready for laser annealing is placed on a processing platform. The angled glass directly behind the wafer is a mirror that is precisely positioned to direct the laser beams to the area to be annealed. (Courtesy Bell Laboratories)

base of the circuit. Then they are coated with a light-sensitive plastic that hardens when exposed to ultraviolet light. After exposure to ultraviolet light, the unexposed still-soft plastic that was masked by a superfine stencil is washed off with acids, leaving behind a hardened microscopic and complex maze that becomes the skeleton of the chip. This is followed by masking, implanting silicon as well as chemical impurities (dopants), and baking. Finally, after a check for performance, each wafer is diced into three or four hundred individual square chips and each chip has silk-thin wires soldered to its input terminals. The chip is then sealed inside a small ceramic case with plug-in prongs for easy connection.

The complexity and detail of the chip are a marvel of microtechnology. Although smaller than a postage stamp, a chip may contain up to a million individual components, mostly transistors. Its appearance under a microscope is very similar to an aerial photograph of a big city, packed with blocks and streets. Electrical pulses carrying coded instructions or information for storage travel the streets of this electronic metropolis almost at the speed of light. During their zigzag journey, which, although erratic in appearance, is well planned and constantly controlled, they perform an amazing number of functions. They turn transistor switches on and off, storing information, amplifying a signal, carrying out mathematical calculations, and even operating machines.

During the last few years the chip has evolved at an amazing speed, growing in "brainpower" and versatility. Its applications are everywhere, and its expected use in many other fields is generating excitement and hope. It has literally invaded our bodies. Inserted next to the heart as a pacemaker, it has become the guardian angel for hundreds of thousands of people with heart problems. Paralyzed people will one day walk again; the deaf and blind will have their senses restored by miracle microchips that can be implanted beneath the skin. From monitoring the weather or the milk yield of cows to booking airline tickets and maintaining proper car fuel injection, the microchip has become an intricate part of almost every electrical device around. To list its many applications would be impossible. It is easier to list the areas that are still untouched by its use, although not for long.

In most of these applications, the microchip is really the heart of a tiny computer designed to carry out a specific task. To many people computers are bulky machines that are instructed with a keyboard or tapes and are often connected to a printer or TV monitor. Few people realize that a digital watch is really a fully functioning computer. It contains all the essential components of electronic activity as a large computer: a central processing unit (CPU), an arithmetic logic unit (ALU), a read-only memory, and random-access memory units.

The digital watch replaced the old-fashioned mechanical timer because of better accuracy and reliability and lower cost. In essence, both watches function very much alike but use different

Three microprocessor chips shown on a wafer. Each chip has more computing power than a minicomputer. (Courtesy Bell Laboratories)

components. Instead of a coiled spring, the digital watch is powered by a battery. In place of the swinging balance wheel, a quartz crystal is set to vibrate 32,768 times per second by the battery. Gears are used in the mechanical watch to reduce the swing period of the balance wheel to one second. In the digital watch, this is done electrically by a series of circuits on a silicon chip made up of almost five thousand transistors. Finally, instead of hands the digital watch displays time in numbers that are formed by the electrical alignment of millions of microscopic liquid crystals.

The microchip is currently appreciated by millions of people with home computers. But indirectly everyone is benefiting from its services. Whether it is preparing a sales receipt, a utility bill, a paycheck, an airline ticket, or in medicine, agriculture, or telecommunications, somewhere a computer is "buzzing" with activity. A faithful and reliable servant, it performs millions of tedious tasks every second without ever complaining. It files information and attacks a problem with infinite variations, while simulating events and forecasting anything from the strength of a designed structure to weather, traffic, industrial efficiency, and even commodity prices. By the end of the century, home computers will be as indispensable as telephones. They will no longer be just home entertainment, another calculator, or a mere filing system. Together with telephones, lasers, and satellites, they will form a powerful technological alliance that will bring about a new information age of unprecedented and inconceivable dimensions. Computers will multiply and diversify. Yet the essential "computer physiology" will be common to all.

There are four major components to any computer: the input, the CPU, the memory, and the output. The lifeblood of the computer is electricity, which flows through its main body in pulses. Every letter or number fed into the computer is first converted at the input into a combination of a low-voltage signal representing 0 and a high-voltage signal representing 1. Since only these two digits are used, the system is known as binary, and the electrical pulses are called bits, from "binary digits." Like the dot and dash in Morse code, bits can make up any letter or number. Most computers use a string of eight bits, known as a byte, for each single piece of information. There are 256 different ways by which 0 and 1 can be arranged to fill a string of eight empty boxes. This is more than enough to express all letters, numbers, and punctuation marks needed for any text. For example, Hello! written in bits is: 01001000 01100101 01101100 01101100 01101111 00100001.

Punch cards, magnetic tapes, or a keyboard resembling a typewriter can be used to deliver information to the computer. In the keyboard system, the most widely used method, each time a key is pressed a string of eight electrical pulses made up of 0 and 1 (byte) is generated and fed into the central processing unit. But raw information is useless by itself. The computer has to know what to do with it. The set of instructions given to the computer as to how to utilize each piece of information is known as a program.

Once a program is introduced and stored in the computer memory, it automatically guides each piece of information to the appropriate place. It is as if cars entering a city were directed by flashing detour signs and forced to park at special parking lots. The detour signs are switched on and off appropriately to pick out individual cars. But both the program and the input information can make sense only if they are written in a special language that the computer can understand. Some computer languages are BASIC, PASCAL, FORTRAN, and LOGO. Each is best for special situations.

The "brain" of the computer is the CPU, which is essentially a high-speed calculator. When electric pulses representing information enter the CPU, they are first read and identified and then directed by the program to different parts of the computer for processing or storage—all at almost the speed of light. To ensure that the signals do not get lost or scrambled during their fast journey through the electronic micromaze, an instruction unit that is part of the chip is employed. It interprets the program and breaks it down to simple steps that are sent for execution to the arithmetic logic unit (ALU), located elsewhere in the chip. For example, most mathematical calculations, no matter how complex, are reduced to either addition or subtraction. In a typical multiplication like 814 × 12, the computer adds the number 814 to itself 12 times. So in essence the computer performs 12 additions, a task that takes less than one hundreth of a second. When two stored numbers are to be added, a code in the calculation program switches certain circuits so that the numbers are extracted from storage and fed together, pulse by pulse, into the adder circuit. There they are combined to produce an output train of pulses representing the answer. As they are, these pulses are meaningless to us. The output converts these electric pulses back into information that we can comprehend. It is given out in a variety of ways, such as a printout or a screen display.

The dramatic growth of the computer industry has brought yet another change, this time in the domain of the English language. The many components and functions of computers have generated a considerable number of new words and acronyms that could fill a small dictionary. "Floppy disks" are plastic disks that carry information, with a program or raw data magnetically inscribed on them. Floppy disks are read inside a small box that operates much like a tape player, known as the disk drive. RAM (random-access memory) is the storage facility of a chip, in which information can be retrieved or stored temporarily. K is the unit of storage capacity of a chip memory and is equal to 1,024 units of information. The famous 64K chip can store 65,536 individual units of information. Undoubtedly the list of computer terminology will continue to grow at an even faster rate, with the expanding computer industry.

To many computer gurus of Silicon Valley, the former orchards in California now packed with computer firms, the current feats of the silicon chip are only a modest beginning. Indeed, the future holds the promise of an infinite number of possibilities. Chips are already the "brains" of industrial robots, and soon they will be able to engage in conversation, diagnose diseases, and even think for themselves. As time goes on, new and more powerful generations of computers will emerge. Electronic engineers plan to squeeze ten million transistors onto a single chip. Eventually, it is

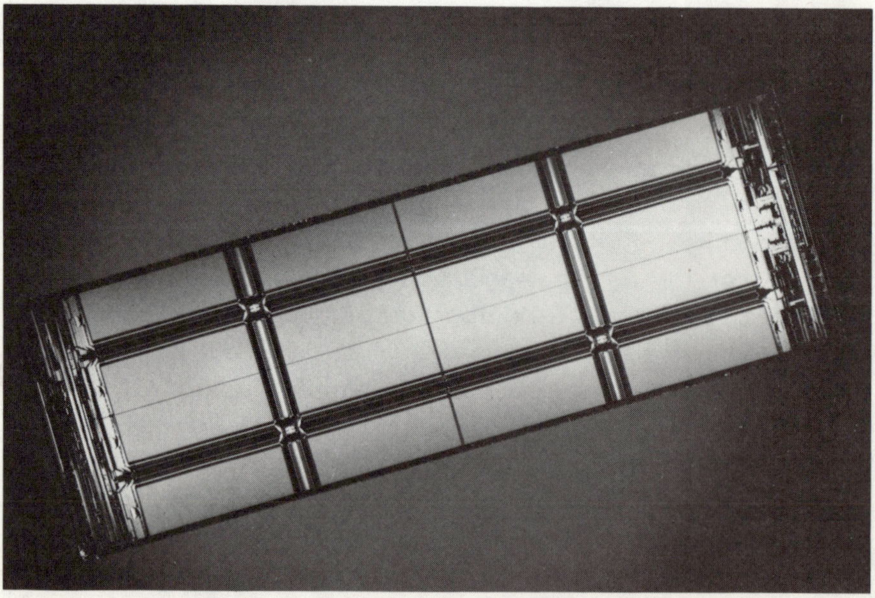

The new megabit computer memory chip of AT&T, which can store over one million digital bits of information. (Courtesy Bell Laboratories)

estimated, a single chip may accommodate up to a billion transistors, enough to store the entire contents of the *Encyclopedia Britannica.*

Supercooling chips in liquid helium can enhance their capacity a thousandfold. But perhaps the most revolutionary computer of the future will contain a chip that is made not of silicon but synthetic proteins, whose molecules may be the ultimate in electronic switches. Serious research is already under way to explore the prospects of this organic chip, the "biochip."

The Information Revolution

A precondition for the development of civilizations was the development of an effective system of mass communications. Without the means to exchange ideas and experiences, humans would have remained primitive, carrying a valuable but unused tool, the human brain. Aside from rudimentary information collected through the senses, such an unused brain resembles a powerful computer without program and data.

History shows that at the very heart of major growth periods in civilizations lie important developments in mass communications. Although often not spectacular in themselves, these developments become essential means by which information can be disseminated widely, a critical prerequisite for advancement. For example, the development of language brought about some direct benefits to primitive humans. Hunting and group defense became more efficient, and social groups grew stronger. First in the form of pictures drawn on the walls of caves and later in the form of letters, written language became the means by which experiences and traditions could pass from generation to generation. But beyond these direct benefits, language became the means by which ideas were articulated to become philosophy, art, religion, and science.

Many key discoveries in history have made it possible to disseminate knowledge more orderly, faster, and cheaper. Once regarded as the privileged property of elite groups such as priests, philosophers, scientists, and artists, knowledge has become more democratic and more readily available. The list of these discoveries is long and includes many that we take for granted, such as paper, ink, pencils, and printing. The invention of the camera and, in 1816, of silver chloride paper (film) enabled people to capture everything that they could see. The introduction of the first newspapers in the sixteenth century, the development of the typewriter in 1829 and of photocopying in 1948, are a few other innovations that spread printed information widely.

The last century has witnessed a spectacular avalanche of discoveries triggered by the harnessing of electricity. Many of these were very important in the field of communications. Morse's telegraph, Bell's telephone, and Marconi's radio system have introduced a global dimension to telecommunications. Despite their diverse natures, they have with their modern cousins television, satellites, and computers the common bond of electricity. Without it none would have been possible. Modern telecommunications therefore owes much to the early pioneers of electricity, like Volta and Faraday. They can justifiably be called the grandparents of our information age.

Systematic scientific research began in the second half of the eighteenth century. Soon after that the volume of recorded knowledge grew to such an extent that it was impossible for one person to read, understand, and remember all that was generated. The solution to this problem was specialization. This in turn fueled the production of even more information and knowledge. Lost in the sea of new writings, scholars were forced to develop a new field—information science—with which information on any subject could easily be retrieved. The Dewey classification system that was introduced in 1876 subdivided all knowledge into ten numbered classes. For example, 100 is philosophy, 200 religion, 500 natural science, and 900 history. With further improvements, the Dewey system became the standard system used by libraries everywhere.

But information and knowledge have continued to grow at a faster and faster rate, not just in the sciences but in many other areas, such as the arts, economics, politics, and law. The number of books and magazines published each year is constantly growing. Too much "undigested" information can cause many problems. One of the pseudosolutions is to continue collecting newer information, leaving behind vast amounts of raw and unprocessed data. The space program, for example, by now has collected enough data to keep research scientists busy for the next twenty years. Yet the foreign competition and the politics of science and economics exert a tremendous pressure to seek the latest. The other pseudosolution is the creation of superspecialists who, buried as they are in their narrow specialization, could be experts in their field but dangerously ignorant and myopic in their peripheral areas. Another major problem is that more and more people know less and less of the knowledge that grows each year by leaps and bounds. The solution to this last problem, which is critical, must be twofold: to make all information readily available and also to make it comprehensible to all. With the discovery of high technology headed by computers, this idealistic solution is slowly becoming a reality.

Computers will be the heart of the information revolution that is brewing right now. Their low cost and many potential uses in communications and information retrieval will ensure their presence in every household. They will prove to be indispensable tools, serving their owners with entertainment, education, health care, communications, and even house security. By themselves they will have limited "brainpower." But connected with giant computer banks, they could retrieve a vast amount of information and utilize many computerized services. By that time the telephone lines making these connections will be modernized also, upgraded for higher efficiency and reliability. They will no longer use copper wires transmitting electric pulses but glass fibers carrying messages in rays of laser light. Branches of these lines could take, if necessary, the signals toward a transmitter, whose satellite (receptor) would ensure worldwide communications. Already newspapers, like the *Wall Street Journal*, are printed simultaneously in many parts of the country thanks to the laser-computer-satellite partnership.

If education is power and information is wealth then twenty-first-century people will be both powerful and wealthy. Not necessarily wiser, although one hopes so. It is difficult to imagine the feeling of having all information that was ever produced available at the push of a button. Books, journals, and even paintings will all be retrievable from a central computer bank, displayed on the monitor

The first of the "wide-body" satellites, LEASATs are designed to take maximum advantage of the space shuttle's capabilities. Each is fourteen feet in diameter and twenty feet high with antennas deployed. Carried in the shuttle's fifteen-foot diameter payload bay, they are placed in orbit to relay telecommunications. (Courtesy Hughes Aircraft Company)

screen or reproduced by the printer. But first all of these will have to be translated into bits and bytes that the central computer can comprehend and store. This mammoth task resembles another period in history, during which the Greek heritage was discovered in the Arab texts. During the twelfth and thirteenth centuries European scholars developed a frenzy of excitement translating Arabic to Latin and resurrecting the great minds of Aristotle, Ptolemy, and Archimedes. These translations became the initial thrust that led Europe to the Renaissance. Translating all our texts to a computer language and making them available to all may spark off a second Renaissance in the twenty-first century.

But education will be but one of the many contributions of the computers. Through a series of questions they will be able to diagnose diseases and report on the general state of health. They may even monitor vital signs and flash out warnings of any upcoming medical problems. Computers will also be the greatest source of entertainment. The film industry is already experimenting in converting films into computer vernacular. One day, all films will be part of a central computer memory and, of course, accessible through a pushbutton. Computerized shopping at leisure from an armchair will provide a large choice of shops, pointing out bargains and quality characteristics of individual products. Even home real estate sales will be computerized. Through computers the layout, structure, area, even shrubs and patios will all be displayed on the monitor screen. Banking, the stock market, commodities, and economic forecasts will be added soon to home computer channels. Most of these services are already part of the successful Telidon system in Canada, serving millions of homes.

All these changes will generate shock waves penetrating every facet of our social life. Consider, for example, the impact of computers on work habits. By the year 2000, a sizable portion of the work force will transmit its work from home via electronic mail. Commuting to work, the nine-to-five workday, sick leave, and vacation time will all be changing. But we cannot yet predict what the indirect social changes will be. Will this signal the end of urbanization? Will big cities begin to shrink, with more people living in the countryside and using the computer and the satellite dish as their social umbilical cord? Will more people staying at home and fewer commuters on the roads mean fewer traffic fatalities, more energy conservation, and stronger family units? No one can predict with accuracy the kind and degree of these social changes brought about by the new information technology.

Like any innovation, computers will bring along also their Pandora's box. People fear the reality of Orwell's *1984*, in which the state, or Big Brother, will rob us of our identity and individualism. There is some justification to these fears. Every time we use computer banks we leave behind tracks that others can follow to make up an image of our lifestyle. Indeed, we expose ourselves to others even now every time we use credit cards. Our habits may not remain private under such conditions. We will be under constant surveillance by banks, governments, credit bureaus, magazine publishers, and every merchant who has interest in our shopping habits. Will federal laws be established in time to guarantee our privacy?

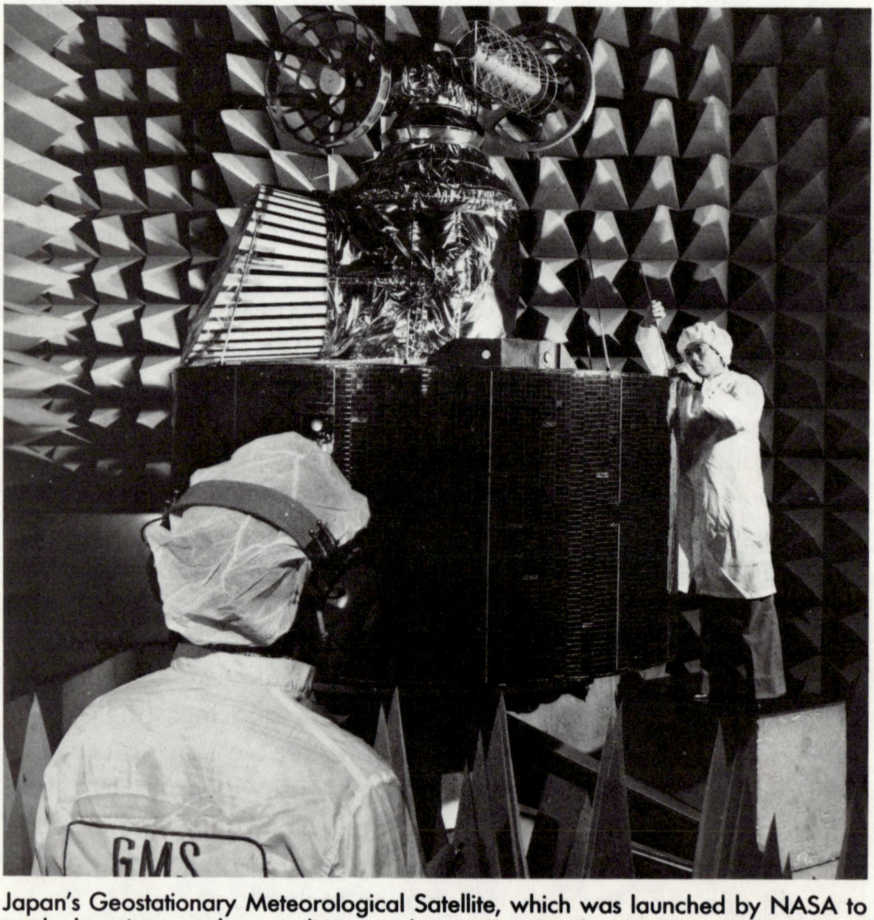

Japan's Geostationary Meteorological Satellite, which was launched by NASA to track changing weather conditions in the western Pacific, shown undergoing final chamber tests. The striped black-and-white unit on the satellite is a sun shield to protect a high-resolution scanning instrument. (Courtesy Hughes Aircraft Company)

Another sensitive legal issue is computer crime. For all kinds of reasons, "pirates" enter computer banks via telephone, and either interfere with, steal, or destroy valuable information. Banking, too, gets its share of high-technology crime. Each day computers transfer billions of dollars from bank to bank. Access to and interference with those computer signals could theoretically transfer billions to an account anywhere in the world. No one knows for sure how much money is lost this way each year. Conservative estimates put it in the many millions of dollars.

The information age has also given birth to a new global awareness that grows stronger as time goes on. News and cultural events are brought right into our living rooms as they happen. Computers one day may become interpreters, able instantly to translate any foreign language into our own. A global awareness and understanding can bring down those artificial barriers of nationality and remove fears and suspicions that are often the result of ignorance. It can also generate in all of us positive responses to injustice and suffering. The horror of the Vietnam War, brought by the media into our homes, generated strong social distaste and disapproval that speeded up its end. The power of the media is most evident in American politics, particularly during presidential elections. This brings up interesting questions. Who are these individuals who actually control the mass media? What guidelines do they follow? Who, if anyone, puts pressure on them to either transmit or withhold a segment of the news? Questions such as these point to the ugly side of the information age. Indoctrination and propaganda have always been instruments of dictatorships. Therefore, along with its many benefits, the information age brings the danger of mass mind enslavement of unparalleled magnitude.

The future is filled with excitement and promise but also uncertainty and fear. Some believe that too much technology is forced on us, allowing us little time to adjust to it. This creates a "future shock" leading to pessimism and a negative attitude toward the future. Others see a bright future in which, with the aid of computers, we will be liberated from many labors. More leisure time will allow people to seek more education and self-enrichment. Others still see the computer as a panacea for all our problems. Such extravagant euphoria is unrealistic. No one doubts the many

Communications satellites have strengthened global awareness and have become instruments of unity and development for many countries. Shown here is an artist's concept of Palapa-B, a communications satellite for Indonesia. Its area of coverage includes Indonesia, Malaysia, the Philippines, Singapore, and Papua New Guinea. (Courtesy Hughes Aircraft Company)

From its synchronous orbit 22,300 miles above the equator, AUSSAT, Australia's first national communications satellite, will link the entire country and Papua New Guinea. (Courtesy Hughes Aircraft Company)

benefits that computers will bring, but even these miracles of technology have their limitations. They cannot satisfy our needs for love, compassion, and social contact. Indeed, this is the sterility of all high technology, not just computers. In an age that becomes more and more technological, it is therefore essential that we ensure the cultivation of these feelings. Otherwise, we may lose our human nature and become machines. Albert Einstein spoke passionately about this need. In a letter to the *New York Times*, published on October 5, 1952, Einstein wrote on education:

> It is not enough to teach man a specialty. Through it he may become a kind of useful machine but not a harmoniously developed personality. It is essential that the student acquire an understanding of and a lively feeling for values. He must acquire a vivid sense of the beautiful and of the morally good. Otherwise he—with his specialized knowledge—more closely resembles a well-trained dog than a harmoniously developed person. He must learn to understand the motives of human beings, their illusions, and their sufferings in order to acquire a proper relationship to individual fellow-men and to the community.

Chapter Five

Medical Technology and Genetic Engineering

A coiled snake, the symbol of healing for many ancient cultures, has survived through the centuries and remains the universal symbol of medicine. It is engraved on Assyrian medicinal cups dating back to 2000 B.C. Moses erected a bronze serpent in the desert, and at this image Hebrews bitten by snakes were to reiterate their faith and be healed. In ancient Greek medicine serpents were regarded as givers of health. Greeks recovering in the temples of Asclepius induced serpents to lick their wounds. Often they even ate snake meat to acquire good health and prolonged life.

There are many reasons why snakes were perceived in that light. Since antiquity they have been considered creatures of acute intelligence and wisdom. According to the Bible, it was the trickery of a serpent that caused the fall of Adam and Eve. Their "earthy" home is the source of many curative substances and medicinal plants. The regular shedding of their skin was proof of their power to regenerate.

If medicine were to seek a new and modern symbol, it would be one that represents science and technology. Any one of their many contributions would be a good choice. It could very well be the microscope, whose window to the microcosmos helped develop the germ theory. It could be the atom, representing nuclear medicine, X-rays, and modern body scanners. Or maybe the laser, whose miracle light has become the most versatile tool in the hands of surgeons. Computers or the silicon chip could also lay claim to the symbol of modern medicine. Tomography, a slice-by-slice view of the body, would have been impossible without them. Another choice could be the DNA molecule, the building block of all life. Its complex structure holds the secrets of unimaginable riches and mind-boggling powers: the cure of all diseases, the slowdown of the aging process, and solutions to global energy and food shortages.

The benefits brought to medicine by science and technology are indisputably many and impressive. But along with these blessings each new medical innovation generates questions of ethics, some of them very controversial, that relate to the legal, social, moral, and even financial aspects of that innovation. Life-support machines, test-tube babies, artificial organs, genetic engineering, abortion, and euthanasia are viewed by some as of great service to mankind, yet others see them as disturbing, unnatural, and even sinful dimensions to medicine.

Most of us consider these innovations inevitable steps in the progress of medicine and the life sciences. Yet we also recognize that they have indirect and often unpredictable influences on our lifestyles and value systems. For issues of life and death we may have to redefine old standards, and for new areas, such as genetic engineering, establish brand-new ones. The development of such standards will require time, a public response, and a wide input from scientists, physicians, philosophers, sociologists, and even lawyers and the clergy.

From Demons to Bacteria: The Development of Medicine

In ancient times illness and death were explained in terms of demons that entered the human body. Treatment therefore consisted of removing these demons by such methods as exorcism, incantations, amulets, and potions noxious enough to shock the patient. Some primitive cultures performed surgery to remove spirits, particularly those thought to cause mental illness or migraine headaches. With a rough stone or a blunt instrument a hole was drilled in the patient's skull, in an excruciating and often fatal operation known as trephining. It was thought that the hole would act as an exit, allowing the trapped spirit to escape. Surprisingly, a few survived such operations, as shown by the growth of new bone found around the holes of some trephined skulls.

The medicine man of primitive cultures was a shaman, a witch doctor–priest–physician in a powerful social position. Dressed in animal skins to embody their power and fearsome looks, the shaman danced around the patient, making noises and ritual gestures to frighten the evil spirits. Although some progress was made in early civilizations, medicine remained intimately linked with religion. In China the concept of the two opposites, yin and yang, dominated many aspects of Chinese life, like medicine, philosophy, and religion. Good health was achieved through the proper balance in the body of all the opposites, which were either yin or yang. Illness was caused by the inbalance of yin and yang, and death occurred when they ceased to flow.

Medical papyri, some dating back to the seventeenth century B.C., show that although Egyptian doctors recognized many diseases and the risks of contaminated foods, they kept medical treatment as a religious practice. Drugs were administered in a religious ceremony filled with chanting. In Assyria, priests dressed in fish costumes sprinkled holy water over the patient while reciting medical prayers. The outcome of the disease was forecast by the position of the stars, oil drops in water, or the entrails of a sheep.

The Hebrews were the first people to practice quarantine, isolation of contagious cases. They became experts in sanitation and public health and were very particular about the nature and preparation of their food. But all these instructions were communicated to them not in the language of science but in that of religion. Such prophylactic measures were considered religious obligations. Sickness was regarded as punishment from God for committing a sin. Repentance and faith in God were necessary elements of recovery. Priests acted as health inspectors and were in charge of allowing recovered patients back to normal social life.

The Greeks were the first people to view medicine rationally. Hippocrates (460–377 B.C.) taught that disease is a natural process that develops in logical and predictable steps. He was the first doctor to keep medical records, noting accurately the progress of the disease, even when it was expected to lead to certain death. To Hippocrates, therefore, the doctor was a man of science and not a priest. The patient was treated as a whole entity, both body and mind, rather than only the individual ailing parts. The famous Hippocratic oath, which has become the standard code of ethics for the medical profession, demonstrates the emphasis Hippocrates placed on responsibility, confidentiality, and trust as essential qualities in the doctor-patient relationship.

Unfortunately, this rational outlook did not last. Soon after Hippocrates and until the seventeenth century, medicine was more of a pseudoscience that used magic and superstition. Many believed in beasts and demons, and medical treatment was often accompanied by faith healing and astrology. Christianity had both a positive and a negative impact on the progress of medicine. The parable of the Good Samaritan developed into a Christian obligation of unselfish and nondiscriminating care for all who suffered. Monasteries in Europe became the first hospitals and centers of intense medical studies. Monks and medical students studied anatomy by dissecting animals such as monkeys and pigs, and also cultivated many varieties of herbs and other medicinal plants.

Unfortunately, Christianity also caused its share of setbacks in the progress of medicine. The belief that on Judgment Day all bodies would be resurrected made the human body sacred, even after death, and its dissection a forbidden practice. What little human anatomy was known was inherited from old texts and was crude and often incorrect. Faith in God's healing powers was established as the prime factor in recovery. Such Christian thinkers as St. Augustine ascribed all diseases to demons. Devout Christians sought cures in prayer and offerings to the Church. Such thoughts and practices did not encourage any serious studies in the cause and treatment of diseases.

In 1543 Andreas Vesalius (1514–64) published *De Fabrica*, a classic book on anatomy, in which the human body was described accurately and in great detail. Soon it became a standard textbook in all European medical schools. Medical students learned anatomy by actually dissecting cadavers rather than observing a classroom demonstration, as had been the tradition for many years.

The intellectual climate of Europe was filled with many similar rebellions in other areas of science. The old authorities were challenged by a new scientific method that employed new investigative tools, experimentation, and measurements. In that new spirit of measurement, the Englishman William Harvey (1578–1657) calculated the rate of the blood pumped by the heart and proved that its motion was circulatory. The human body was no longer considered mysterious but a beautiful and complex machine obeying the same laws of science that nature did.

Medicine became the true science that it is today at a rather slow pace during most of the last two hundred years. Forming alliances with other sciences, such as chemistry and physics, and making use of such new tools as the microscope and X-rays, medicine has become an effective, exciting, and respectable discipline. During these last two hundred years major discoveries were made that saved millions of lives.

One such discovery was made by the English country doctor Edward Jenner (1749–1823) at the end of the eighteenth century. Jenner was puzzled by an observation he made, that dairymaids who contracted the mild disease of cowpox never caught smallpox, one of the most dreaded diseases then known. He concluded that the cowpox helped develop immune resistance to smallpox. Ironically, a British antivaccination society was formed to arouse public fears on the danges of vaccination. A popular 1802 cartoon showed people who were vaccinated as having little cows growing out of their bodies. Despite these malicious attacks, Jenner managed to spread the practice of vaccination, supported by a grant from the British government.

Another major breakthrough in medicine was the discovery of the connection between disease and germs, or bacteria. The most convincing experiments to support this connection, known as the germ theory, were carried out by the French chemist Louis Pasteur (1822–85). Through his investigation of the process of wine fermentation, Pasteur showed that germs are not born spontaneously, as many thought at the time, and that they are present in the air we breathe. He showed that the transfer of germs by air could be prevented by filtering the air through cotton, wool, or mineral granules.

Pasteur went on to develop a method for preventing wine or milk from becoming sour. The process, known as pasteurization, involved heating the liquid for a few seconds at 60°C (140°F) and is widely used today in the milk industry. Pasteur recognized also that germs that were heat-treated could be used for inoculation to strengthen the immune system.

The German bacteriologist Robert Koch (1843–1910) identified many disease-causing bacteria, like those of cholera and tuberculosis, and developed a number of laboratory techniques still in use today.

The study of bacteria would have been impossible without the microscope. Developed during the seventeenth century, it was originally used almost as a toy, a mere window to a new and fascinating world. The first microscopists even failed to make any association between the tiny creatures and disease. When that association was finally made the microscope was recognized as an indispensable medical tool. Other such tools were the clinical thermometer, introduced in 1866 by a Briton, Clifford Allbutt (1836–1925); the stethoscope, which was invented in 1816 by René Laënnec (1781–1826), a Frenchman; and, of course, the X-ray tube, invented accidentally by Roentgen in 1896.

In surgery the two most important events, both in the mid-nineteenth century, were the development and application of anesthetics and antiseptics. Prior to those discoveries surgery was primitive and very painful, with a low survival rate. Operations, mostly amputations and removal of external tumors, were carried out as a last resort.

In 1800 Humphry Davy, England's most famous chemist, proposed that laughing gas (nitrous oxide) could be used as an anesthetic. His proposal was ignored for almost fifty years. In 1846 William Morton (1819–68) used ether at the Massachusetts General Hospital, and a year later James Simpson (1811–70), a Scottish obstetrician, used chloroform as an anesthetic and proposed its use in childbirth. The Scottish church attacked such a practice on theological grounds, basing the arguments on a passage from the Bible that women should have pain in delivery. Fortunately, Queen Victoria granted anesthesia the royal stamp of approval by using it herself in two of her confinements.

Although anesthesia took away most of the pain in surgery, it could not make postoperative survival any better. The real killer after surgery was infection, or sepsis. Doctors operated in their everyday clothes, without sterilizing their hands or cutting instruments. Some even smoked cigars during operations. The dirty environment of hospitals did not help, either. Infection by contact was first suspected by a Hungarian obstetrician, Ignaz Semmelweis (1818–65). In 1847 he instructed doctors and medical students under him at the Vienna hospital to wash their hands in strong chemicals and to scrub their fingernails with a nailbrush before entering the maternity ward. The result of this simple precaution was the dramatic decrease of the mortality rate of pregnant women from 12.4 percent to 1.3 percent. Unfortunately his colleagues laughed at the simplicity of his precautions. In 1867 Joseph Lister (1827–1912) published the results of his investigation into mortality rates. He argued that the use of such chemical antiseptics as carbolic acid on wounds and instruments and sprayed throughout the operating theater could dramatically increase the survival rate, as indeed was the result in many of his own operations.

Medicine lists many spectacular discoveries in the twentieth century. The roles of hormones and vitamins were identified. Viruses were discovered; the mechanism of the transmission of nerve signals was explained; the chemical code of life, or DNA, was described, and the role of genes in heredity was identified. New wonder drugs were discovered, accidentally or by design. In 1928 Alexander Fleming (1881–1955) discovered penicillin, which saved millions of lives during World War II alone. Smallpox was eradicated from the face of the earth, and a polio vaccine was discovered in the 1950s by Jonas Salk (1914–).

As time goes on, medicine is becoming more and more dependent on the high technology of such sciences as physics and chemistry. Major discoveries are often adapted to medical needs, providing diagnosis and treatment of what were once considered incurable conditions. Hospitals have evolved to resemble space-age laboratories, furnished with sophisticated equipment like CAT scanners, lasers, computers, particle beams, ultrasonic blasters, and thermographs. The number of scientists employed by hospitals to operate these new tools has grown steadily over the years, another testimony to the new technological face of medicine.

Medical Technology

Since their discovery in 1896, X-rays have served medicine faithfully. In their early applications they were used only to detect fractures and such foreign bodies as lead bullets. Gradually their use in medical diagnosis expanded, and today X-rays are used to detect cancers, ulcers, a number of diseases, and, with the help of injected tracers, problematic organs or body functions. Yet despite their sophistication and extensive use, their ghostly film images provide only limited information about the internal structure of the human body. Furthermore, their dangerous biological effects restrict their use.

Since the early 1970s three revolutionary methods have been developed in diagnostic medicine. They use machines known as CAT, NMR, and PET scanners, which produce three-dimensional images of the human anatomy with unprecedented clarity and detail. Although they have distinct differences, all share common characteristics. For example, they all use a large cylindrical chamber that scans the human body slice by slice, much as a knife slices salami. Detectors rotate around or surround the scanned part of the body. They are connected to powerful computers that use their signals to build an image of that particular cross section of the human body.

The CAT (computerized axial tomography) scanner was invented in 1972 by Godfrey Hounsfield (1919–) and Allan Cormack (1924–), for which they shared the 1979 Nobel Prize for medicine. An X-ray source and a set of detectors on opposite sites move around the body. The signals from the detectors, transmitted X-rays, are fed into the computer, which reconstructs the image

Radiologist Dr. Clyde Williams prepares a patient for NM Resonance scanning. (Photographs in this chapter courtesy University of Florida Health Science Center)

of that particular slice of the body. The X-ray source and detectors are then shifted slightly to another point and repeat the process for the adjacent slice.

One of the advantages of the NMR (nuclear magnetic resonance) scanner is that it does not use X-rays. The human body is exposed instead to a strong magnetic field as it goes through a doughnut-shaped magnet. This forces the nuclei of hydrogen atoms in the body to align themselves like tiny compasses in the same direction as the field. But because nuclei also spin, the net result is that they wobble like tops. Different elements in the body wobble at different frequencies. When a second, smaller oscillating magnetic field is applied at right angles to the first one, it causes some of the nuclei—specifically, those that oscillate in the same frequency as the second field (they are said to be in resonance)—to be realigned. When the small field is turned off they fall back to their original orientation, giving off a characteristic signal. The computer then processes all these signals and displays them as chemical data or as an image.

NMR scanners provide images of the human anatomy that are difficult to see with any more conventional method. They have great clarity and detail and are very sensitive in differentiating between diseased and healthy tissue. Even tiny cancer tumors can

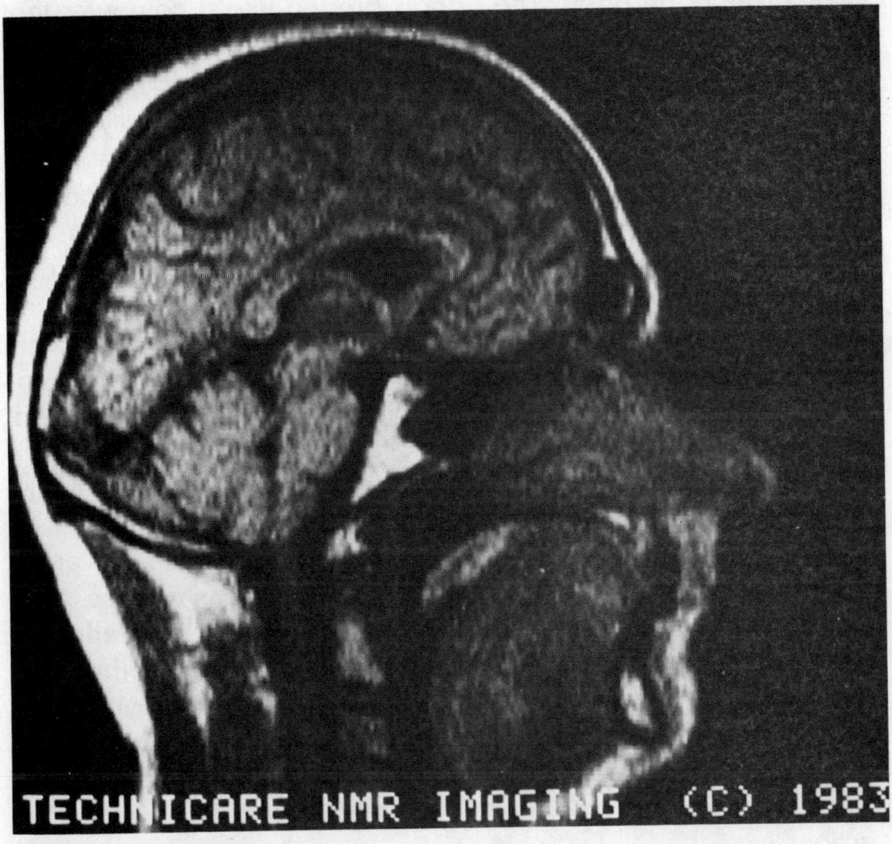

NMR scan of cranium.

be detected. But besides these advantages of clarity and resolution, the NMR method has a unique advantage over the CAT scanner in that it can detect body functions in action. Moving organs, the flow of blood, and the effects on them of certain drugs can now be observed live, as they take place. Since most of the signals come from the hydrogen in watery body fluids, the NMR method cannot be used to study bones. Because the NMR signal is very sensitive, it is necessary to insulate the room magnetically, adding an extra cost to an already very expensive machine.

The PET (positron emission tomography) scanner uses positrons (positive electrons) emitted by radioactive tracers injected into the body. When a positron collides with an electron, two gamma ray photons are produced, which travel outward and in opposite directions. These photons are picked up by detectors placed in a circle around the body. The signals are fed into a computer that can calculate the exact location of their origin inside the body. The computer quickly processes all this information and then draws a picture of the body, slice after slice. The picture is really one of intensity variation, each region represented by either a degree of darkness or a different color. To a radiologist this intensity pattern shows structure as well as chemical and metabolic activities.

Although in theory the PET scanner can study the structure and function of any living organ, it is mainly used for the brain. It is the only instrument that can actually see what goes on inside the brain when sensory signals like sight and sound arrive. This will undoubtedly enrich our understanding of this most powerful organ, of which, by comparison to other organs, we know very little. In these color mappings of the PET scan we can actually see how the brain thinks.

Its valuable images can pinpoint brain damage from epilepsy, schizophrenia, Huntington's disease, and Alzheimer's disease. Such information may one day help arrest their growth through effective drugs. Even the physiological and chemical changes of the brain during anxiety, depression, fear, and euphoria can be observed. It was found, for example, that panic-disorder behavior (sudden attacks of panic) can be linked to lactic acid in the body. More mental illnesses will one day be reduced to specific chemical reactions.

Despite their high cost (up to $1.5 million), each of these scanners is a powerful diagnostic tool, a unique window that looks at the structure and physiology of the human body. Although they are all in their early and still imperfect stages, their amazing results have impressed many hospitals sufficiently to buy them. These magnificent machines are a splendid example of how fruitful the alliance between computers, advanced physics, and medicine can be.

Another new diagnostic tool is thermography, the analysis of invisible heat rays (known as infrared) emitted by the human body. The patient stands naked in front of a detector that scans the body from a distance. The signals are converted into numerical data and fed into a computer that uses them to build an image on the screen. Although the image is that of the patient, it has no detail and from it you cannot identify the individual. Instead, it is made of regions of colors, each representing a different temperature. Doctors look for asymmetries in the color pattern, which are due to hot spots, regions of the body that give out more heat than healthy tissue. These can be from injuries, cuts, burns, or early signs of a major problem, like impaired circulation, that can lead to a stroke, or cancerous tumors, like those found in the breasts.

With so much emphasis on preventive medicine, it is not at all surprising that so many new diagnostic methods have been developed lately. One day we all may be able to diagnose a problem using the phone and conversing not with a doctor but with a central computer. Such computers, like CADUCEUS at the University of Pittsburgh School of Medicine, have already been developed. Programmed by physicians, these computers will be taught how to ask the right questions and in the right order. Constantly consulting their memory, which may contain all the details of thousands of diseases, in minutes they will provide a diagnosis and a list of recommendations.

Soon after their discovery lasers were seen by physicians as a medical tool of great potential. Today they have become one of the most exciting, effective, and versatile tools in diagnostic and curative medicine. Their miracle light is used to identify cancer cells and to weld, drill, or vaporize diseased tissue. Laser beams are carefully focused or guided through fiber optics to a precise point and are fired for a time that, according to the need and the type and power of laser, can be as long as thirty minutes or as short as a billionth of a second.

The most popular and successful medical use of lasers is in eye surgery. Transmitted through the lens to almost every internal part of the eye, light from a variety of lasers can deliver four kinds of energy levels and so perform a number of different operations without the need to cut the eye open. Focused on the retina, a laser beam can seal a hemorrhage or a tear. Producing tiny burns, it can halt the abnormal growth of blood vessels caused by diabetes. It can literally blow up tiny particles floating in the vitreous humor. It can drill a series of tiny holes around a membrane that has grown milky to restore clear sight. It can drill tiny holes around the iris to drain blocked fluid and relieve pressure in glaucoma patients. In all these operations the laser action is instant and virtually painless.

The leading cause of death in the United States is arteriosclerosis, the blocking of the arteries by plaque deposits of fat and calcium. Depending on their location these blockages can cause heart attack, stroke, or gangrene. At least 100,000 Americans each year undergo coronary bypass surgery, the most widely used remedy for severely narrowed arteries, at an average cost of $20,000. Surgeons have turned to the laser for a solution to this problem. The laserscope, a laser device that can chisel away fatty deposits, may be the answer.

The laserscope is a three-tubes-in-one catheter that is inserted into the problematic artery. One tube is made of quartz fibers that carry the laser beam. Another tube, made of optical fibers, transmits ordinary light to illuminate the artery. The third tube sprays fluid to clear the target area and then vacuums away solid debris and vapors. Right behind the tip of the laserscope is a balloon, which is inflated to halt the flow of blood.

Under ideal conditions, a laser-beam catheter threaded into an artery could blast away plaque precisely and without damaging the walls, burning passageways through clogged arteries. Although the laserscope is still at an experimental stage, its inventor, Garrett Lee (1946–), believes that in a few years, after improvements and a lot of practice on animals and cadavers, it will offer

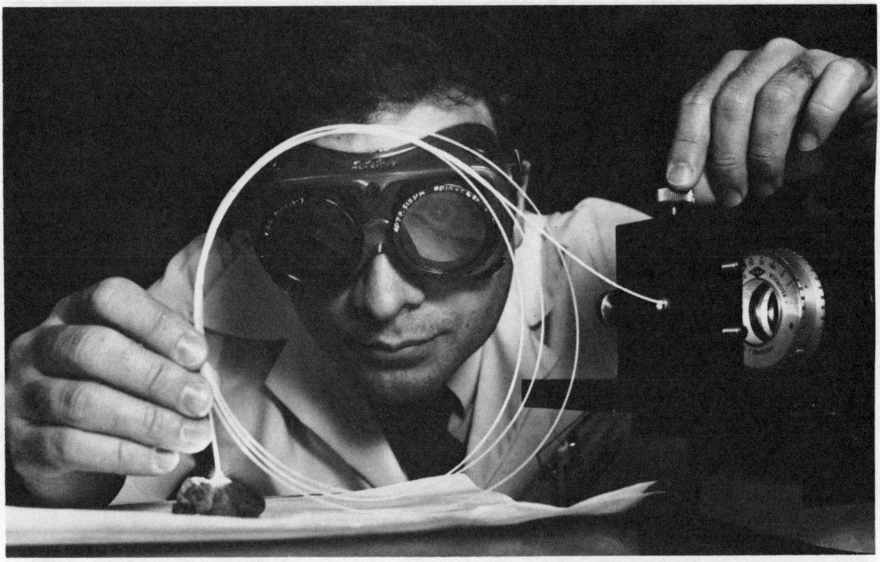

Heart specialist Dr. George Abela tests an argon laser on a rabbit heart. He has successfully used three types of lasers to clear blocked coronary arteries.

a viable cure for a killer disease.

Sound is also being used in medicine, but not as widely as laser light. As ultrasound (not audible by our ears), it can scan the fetus and monitor its growth in the womb. Many women obtain the first picture of their unborn child in this way. Any complications during pregnancy can be spotted with ultrasound scans and without the need for X-rays, which can harm the fetus.

Lately sound has been used in yet another impressive application, the treatment of kidney stones. For the 500,000 Americans who suffer from the excruciating pain of kidney stones each year, two new radical methods have been introduced to replace conventional surgery. In one method the body is immersed in water and subjected to shock waves focused on the kidney stone. In thirty to forty-five minutes and after eight hundred to a thousand shocks, the stone is reduced to gravel, which is finally excreted within two months, causing only minor pain. In the other method, a probe is inserted in the patient's back, right up to the kidney. Through it the surgeon sees the kidney stone and bombards it with sound vibrations for twenty minutes to cause its disintegration. The

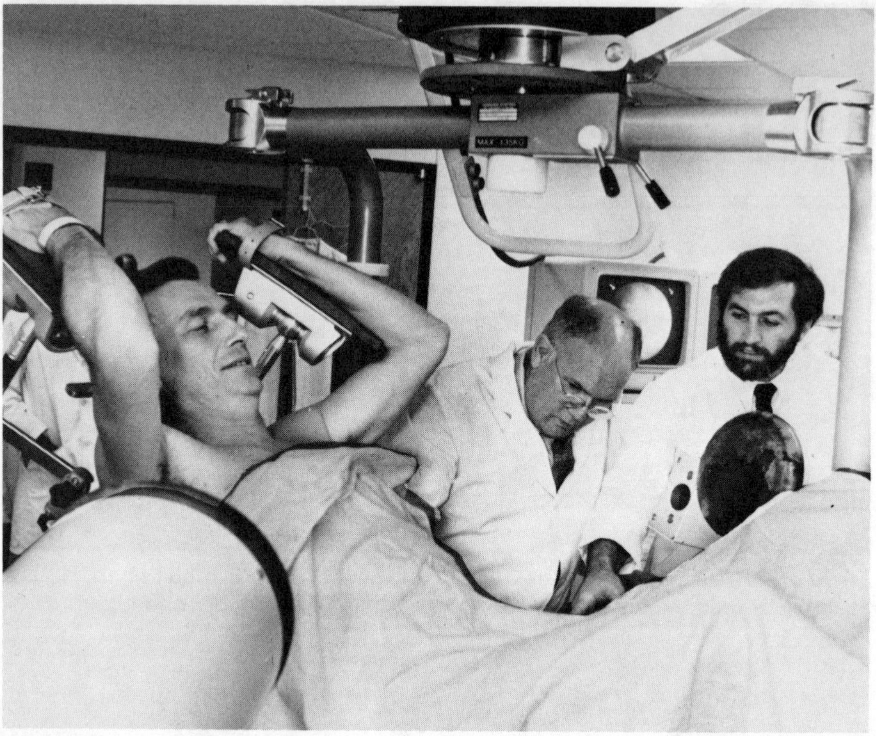

A patient prepares for an underwater shock-wave lithotripsy. Within an hour his kidney stone was disintegrated.

The result of a shock-wave lithotripsy *(right)* on a typical kidney stone *(left)*.

fragments are then flushed out through the probe. Neither method causes any damage to healthy tissue, and both enjoy a fantastic success rate of 90 to 98 percent. Their effectiveness, short recovery period, and lower cost make them both attractive alternatives to surgery.

Surgery without cutting is also being tested in the treatment of brain tumors. The patient is strapped to a chair and wears a fiber glass head-shoulder cast that looks like medieval armor and holds the head in a fixed and precise position. The brain is scanned and the location of the tumor is fed into a focusing device. This directs a beam of subatomic pions precisely at the tumor for a few minutes each session. The pion beam is produced by accelerating a beam of protons to almost the speed of light and then aiming it at a target; the pions are some of the fragments. Pions are used because it is believed that they can cause minute explosions in malignancies, leaving healthy tissue intact. Beams of high-energy neutrons have also been used effectively to destroy brain tumors without surgery.

Although these large machines tend to capture most of the medical headlines, there is also a small-scale biotechnology, just as impressive, in the research and development of spare parts for the human body. The partnership of material sciences and bioengineering has produced hundreds of artificial devices that can replace parts of the body damaged by accident, removed by surgery, or weakened by disease. Some of the most widely used and successful implants have been heart valves, electronic larynxes, hip joints, and prosthetic limbs, which are often packed with springs, levers, and microelectronics for wide mobility. Pacemakers implanted under the flesh feed the heart with minute electric shocks that regulate its contractions. The near future is even brighter for such devices. Microelectronics combined with tiny computer chips will become the basis of many portable or implanted devices that can monitor vital functions such as blood pressure and heartbeat. They will be programmed to identify and warn about irregular biopatterns that can lead to a crisis, such as a stroke or a heart attack.

The most sensational transplant operation, which made medical history on December 2, 1982, was that of the artificial heart. For

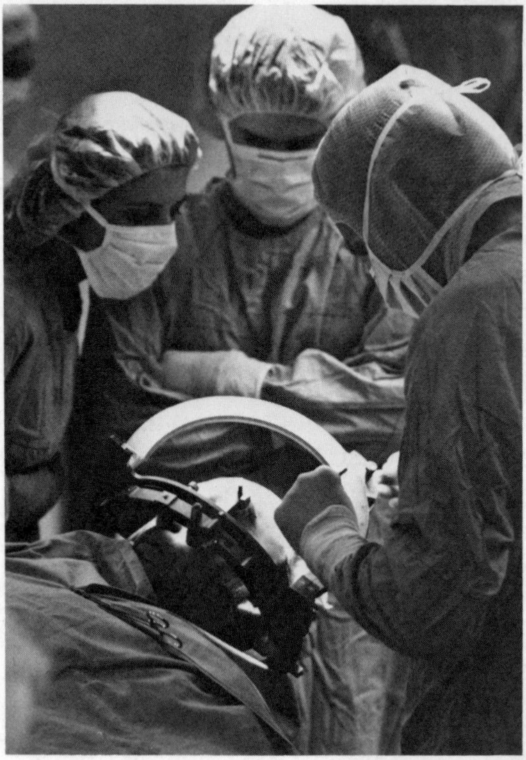

Navigating deep inside the brain with the aid of scanners, computers, and a metal headband, neurosurgeons implant radioactive "seeds" to destroy or slow the growth of inoperable brain tumors.

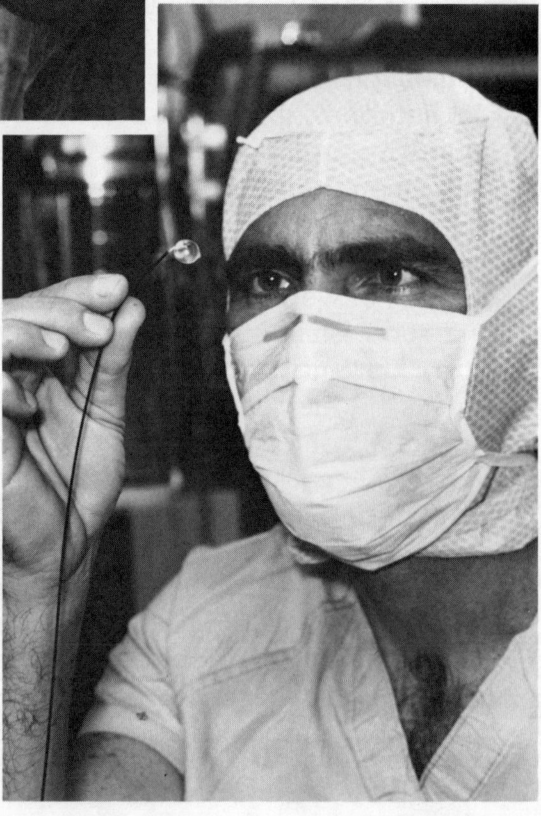

Dr. J. Parker Mickle of the University of Florida inspects the tiny rubber balloon that can spare some patients brain surgery.

the first time in history a human, sixty-one-year-old retired dentist Barney Clark, had a heart replaced by a mechanical gadget. Clark's heart was deteriorating fast, and he had only hours to live. Despite an early valve malfunction that required emergency surgery, the artificial heart kept him alive for 112 days. His death was attributed to lung and kidney failure.

William DeVries (1943–), the young surgeon who performed the operation at the University of Utah in Salt Lake City, used the Jarvik 7 model, the latest design of another young physician, Robert Jarvik (1946–). Jarvik worked for over twelve years with Willem Kolff, the inventor of the kidney dialysis machine and director of the artificial organ program at the University of Utah. In 1957 Kolff and Tetsuzo Akutzu (1922–) developed an artificial heart that kept a dog alive for ninety minutes. Since then Kolff has remained a world authority on mechanical hearts.

Despite its crude appearance, the Jarvik 7 artificial heart has proved itself to be a sturdy and functional life-prolonging gadget. It is made of polyurethane and consists of two independent pumps, each with a flexible diaphragm. An external compressor drives the artificial heart with puffs of air, which are regulated by a machine called the heart driver. Two six-foot-long pipes carry the air to the heart, entering the body below the ribs on the left side of the abdomen.

In the field of material sciences new synthetic compounds are developed and tested by scientists trying to match their properties with those of the human body. Membranes have been developed that can save burn victims, plastics that can make eye lenses, and Dacron tubing that can replace blood vessels. These materials not only have to match their organic equivalents but also must be accepted by the body. Often rejection has been a problem. The chances of this happening to Bioglass are very slim.

Bioglass is the first synthetic material able to bond with living tissue. Its inventor, Dr. Larry Hench (1938–), a ceramic engineering professor at the University of Florida, believes that it will soon be used to replace middle-ear bones for those with impaired hearing. It will also be used as a denture implant and as a coating for artificial hip joints.

As with every new technology, medical technology has brought along with its benefits some serious social and ethical issues. Making early diagnosis of deadly illnesses possibly will result in the cure and survival of thousands who otherwise would die. Life-sustaining machines and gadgets can easily increase that number to millions. But while some see this new technological dimension of medicine as a blessing, others consider it detrimental to our species, sinful, and costly. One of their arguments is that more and more people who would have died in infancy a few years ago survive today to produce children of their own, perpetuating their defects.

It is rather ironic that as medicine has advanced over the years it has become more difficult to define the exact moment when life begins and when it ends. The issue of abortion, which has haunted us for years (Hippocrates forbade his physicians to induce it), has grown more complex and certainly more controversial. The rights of the fetus versus those of the woman have become entangled in medical, theological, and legal battles. Abortion has often become a political issue. Test-tube babies, artificial insemination, and fertility drugs have all generated controversy.

The issue of euthanasia, or mercy killing, is equally controversial. Patients in coma after irreversible brain damage are often kept alive by machines, at great financial cost and emotional anguish to their relatives and doctors. Who decides whether or when the machines should be turned off? The physician, the family, or the courts? In the classic case of Karen Anne Quinlan, the court granted her father permission to act as her guardian. He chose to turn off the mechanical respirator. Unexpectedly, Karen Anne Quinlan continued to live for nine more years in her coma, until her death in July 1985.

For some years now hospital committees have been deciding who lives and who dies on a weekly basis. With limited kidney dialysis machines and transplant organs, hard decisions as to who the recipients will be must be made. But in the case of active euthanasia, neither the physician nor the hospital committee can decide. Such decisions are difficult to make, and it is obvious that each case must be measured on its own merits, requiring input from many concerned parties, even the clergy, whose advice may be sought by the family.

But even without these life-supporting machines, the issue of euthanasia would still be with us. Terminally ill people who can no longer stand the pain may wish to die with dignity and spare their loved ones the cost and anguish. The medical profession must establish guidelines that incorporate ethical, legal, and theological concerns on euthanasia. This may be a difficult task but, as more and more people choose euthanasia, a very necessary and urgent one.

The high cost associated with medical technology is another thorny ethical issue. Is it worth spending many thousands of dollars on a medical intervention that may add only weeks or months of life? Is it fair that only the rich can afford such luxuries as a PET scan? Should this money be spent instead on preventive medicine? Should the criterion for living be life or good life? Maybe our Western culture needs death education and the development of a new philosophy and criteria for life worth living.

Genetic Engineering

A new age in science was launched in 1953 when two scientists, an American biologist and a British physicist, working together at Cambridge University made a brilliant and historic discovery. James Watson (1928–) and Francis Crick (1916–) unraveled the mystery of DNA (deoxyribonucleic acid), the blueprint of life. Using X-ray observations made by Maurice Wilkins (1916–) of London University, they proposed that the structure of DNA is a double helix, shaped like a spiraling ladder. All three shared the 1962 Nobel Prize in physiology (medicine).

Two earlier milestones in the history of biology were the works of Charles Darwin (1809–82) and Gregor Mendel (1822–84). Darwin's classic books *The Origin of Species* (1859) and *The Descent of Man* (1871) placed humans in the animal kingdom, which goes through a continuous evolutionary process controlled by natural selection. Darwin's theory of evolution caused an earthquake in the intellectual climate of the late nineteenth century. The tremors are felt even today.

Mendel, an Austrian monk curious about the mechanism of heredity, began in 1856 to experiment with peas in the monastery gardens. He discovered that there are dominant and recessive characteristics and that it is possible to predict the probability of offspring with specific characteristics. From these simple yet elegant experiments Mendel was able to identify the laws of inheritance. His work was published in 1866 but was virtually ignored until 1900, when its value was recognized. Mendel's work was the foundation of a brand-new discipline—the science of genetics.

To understand inheritance one has to understand the unit of life, the living cell, which functions like a miracle machine. The cell not only has to perform its own duties as part of the living organism but also has to duplicate itself. Its control center, the nucleus, contains coils of DNA, which carry the instructions for all the functions of the cell in coded segments called genes. If the DNA is considered as a spiral ladder, then its sides are formed by phosphates and sugars and its rungs by pairs of four chemical bases, adenine, thymine, guanine, and cytosine. It is the order by which these bases combine that makes up the genetic information in the cell. This is what gives a liver cell its "liverness," a flower its color, and an animal the shape of its nose. Packed like random words in an encyclopedia without an index, the characteristics of each individual organism are locked in the complex molecule of its DNA. Deciphering this complex code and then tinkering with its structure can yield desirable or unpredictable characteristics.

In the 1960s molecular biologists trying to decode the DNA molecule obtained an unexpected clue from their studies of the defense mechanism that bacteria use against viruses. They found that bacteria secrete a substance that splices the DNA of the attacking virus into pieces. The next breakthrough came in 1973, when scientists managed to insert "quest" pieces of DNA into the genes of a normal "host" bacterium. When the bacterium reproduced itself, it reproduced also the quest pieces, as if they were part of a new organism. This process is now called recombinant DNA.

The ramifications of recombinant DNA, this microscopic gene transplant, are mind-boggling. Thousands of medical questions can be asked about what part of DNA causes this or that disease, weakens this or that organ, or even causes aging itself. Then, of course, come questions as to how to reprogram these parts, if possible. Agricultural scientists can ask similar questions, seeking to improve crops and animals, to make them more efficient, increase their yield, and make them resistant to disease. They can envision plants that can make their own fertilizer and citrus trees that cannot freeze. All this was still science fiction twenty years ago, but early results show that all this can one day be achieved.

Some of the first beneficiaries of recombinant DNA have been diabetics. Insulin taken from pigs and cattle to supplement the body's declining production is expensive and can often lead to allergic reactions, and its supply can fluctuate with the livestock market. In 1978 scientists managed to implant human insulin genes in the common bacterium *Escherichia coli.* The implanted bacteria grew quickly into colonies, reproducing also the insulin gene that they carried, which could later be extracted. Today large quantities of safe and inexpensive insulin are biomanufactured and marketed under the trade name Humulin.

Drugs that fight cancer have been developed in the same way. One of them, interferon, regulates the immune system and is produced naturally in the body but only in tiny quantities. Genetically manufactured interferon can stimulate cells called macrophages in the bloodstream to seek and destroy enemy cells like those in cancer and possibly viral infections such as hepatitis and herpes.

Recently there has been excitement about a new bioengineered protein that can help dissolve blood clots that lead to stroke or heart attack..

Children suffering from dwarfism due to a deficiency of a growth hormone produced in the pituitary gland may also be a problem of the past. Genetically manufactured hormones can now allow them to attain a normal growth. Similar growth hormones when injected in livestock can increase their yield dramatically. Tests on cows have shown an increase in their milk production by as much as 30 percent. The list of these miracle hormones will become very long in the next few years. The Food and Drug Administration will require years of testing, however, before approving them for marketing.

Solutions to the global need for food may also be provided by recombinant DNA in a variety of ways. One way is to modify the genetic code of plants such as wheat and corn so that they can make their own fertilizer from atmospheric nitrogen, just as beans do with bacterial colonies in their roots. Another way is by genetically designing plants that can yield more nutritious crops. There are, for example, several plants that can grow in abundance in third world countries but, due to the lack of some amino acids, their crops provide little nutritional value, making them unworthy of cultivation. One day soon the DNA code of these plants will be genetically altered to include the required instructions that can make the crops highly nutritious.

Genetic agriculture is undoubtedly an exciting new science with infinite possibilities and tremendous global potential. In theory, at least, scientists can add any desired quality to the DNA of any plant, like resistance to disease, drought, or frost or increased yield, more nutrition, or better taste. Scientists will one day be able to make a hybrid plant whose parts—roots, leaves, and fruit—can all be of some use. Imagine, for instance, a plant with potatoes as roots, tobacco leaves, and tomatoes as its fruit. Farfetched? Not really.

It already takes place in test-tube gardens. Using a technique called protoplast fusion, scientists have managed to combine the genetic material of two plants into a totally new hybrid, never seen on the face of the earth. Mass production of these new plants is achieved by cloning, in which ordinary cells of the plant are tricked into behaving like embryos, each cell developing into a plant. As many as a thousand plants can be cloned from a single gram of starting cell. The human equivalent of cloning would be to produce carbon copies of people from a piece of skin tissue, a frightening possibility indeed.

Even the energy industry is excited about genetic engineering. In the United States alone millions of tons of cornstalks, sawdust, and other trash have to be disposed of every year. Locked into the molecules of these organic leftovers are valuable sugars that could be converted into alcohol. The problem is cellulose, a protective cell layer that bacteria cannot break through to ferment the sugars into alcohol. A new breed of bioengineered bacteria may one day be able to do that, producing millions of barrels of alcohol that can be mixed with gasoline to make gasohol for cars.

Under consideration is the design of a superbacterium that could free oil left over from drilling and clinging tightly to the rock surfaces. This extraordinary organism would have to survive without oxygen and withstand the high temperature, pressure, and salinity in the depths of oil wells. A very tough task indeed for a genetic engineer, and it may not happen for some years, but if and when such a bacterium is produced the rewards could be as much as 200 billion barrels of oil, worth trillions of dollars, in the United States alone. How about bacteria that eat chemical pollutants, or scavenge metals from low-grade ores? All of these are under serious investigation. Recently a U.S. patent was granted for a manufactured oil-eating bacterium.

It turns out that every time we split nature's most fundamental building blocks we stand at the verge of utopia as well as catastrophe. In 1939 we split the atom, and look where we are today. We have nuclear energy, nuclear medicine, and along with them nuclear waste and the frightening prospect of a nuclear holocaust. In the 1960s we split DNA, and to say that the impact will be ten times that of the atomic age is certainly an underestimation. No past scientific discovery can match the revolution that genetic engineering is about to make.

The commercial side of genetic engineering, a vital force in its growth, was given a boost in 1980, when the U.S. Supreme Court decided that genetically modified bacteria can be patented. In only a year 150 biotechnology companies had been established, along with many genetic-engineering divisions that were set up by chemical and pharmaceutical corporations. Billions of dollars have already been invested in this very lucrative enterprise, which has already convinced many that it will soon outrun the computer industry.

Capitalists competing to capture the biotechnology market are doing their utmost to ensure that the best professors in molecular biology and genetics go on their company's payroll. Even as consultants these academics will have to sacrifice their privilege of freely sharing scientific information. But the temptations are so tantalizing that scientists are already leaving their academic posts to form their own companies. Even universities are considering getting into the business.

Philosophers and educators are alarmed by this situation. They fear that this may cause an irreversible erosion of the academic environment, where information is freely sought and exchanged for knowledge's sake and scientists are neutral, representing all humanity and not special groups whose motivation is profit. Others argue that this symbiotic relationship between universities and corporations is inevitable and even welcome to many universities that are struggling to keep their research going.

Besides the commercial and academic issues, genetic engineering has raised some interesting ethical questions, with many more still to come. What gives us the right to play God and tamper with nature's evolutionary wisdom? Is it in our best interest to cure all the sick people through gene therapy? Will it be wise and ethical to allow the cloning of people for whatever purpose? Do we really comprehend all the dangers involved? How likely is it that a manufactured superbacterium could escape and wipe our own or any other species off the face of the earth?

Questions like these will be intensely debated for many years. Their answers will require the input and wisdom of many professionals besides biologists. Philosophers, sociologists, the clergy, even lawyers and politicians will have to struggle together to reach answers acceptable to the public. Knowing the diversity of opinion in our free society, this may indeed prove to be a Herculean task. In the meantime, guidelines must be established, and strict controls enforced by a federal agency must be observed. We cannot afford to allow mistakes or accidents in bioengineering. Both could be irreversible.

The battle of opinions has already started, occasionally reaching the courthouse. Dedicated environmentalists act like modern Luddites and insist on a complete stop to all work in genetic engineering. To them this new science is like a modern Dr. Frankenstein, whose manufactured microscopic monsters will one day get out of control and bring an end to life as we know it. Others with more faith and optimism argue that we have no right to stop the progress of science. Indeed, if opposition had prevailed every time science brought danger, Prometheus would have never been sent to steal fire from the gods.

Despite the grave dangers and ethical issues posed by genetic engineering, the intellectual and material rewards are too precious to ignore. Genetic engineering must and will be allowed to continue. There is even an extraterrestrial element to it. Manufactured life that can withstand the most peculiar and hostile environments may one day be transplanted to other planets. If such synthetic genesis were ever to be accomplished, would it be a blasphemy to God and nature? Or would it be a proper use of our endowed gift of intelligence and a necessary forward step to our destiny?

Chapter Six
Relativity

The theory of relativity has been called the greatest triumph ever achieved by the disciplined imagination. When it was first proposed, in 1905, it shattered the traditional concepts of space, time, energy, and matter. Twelve years later its extended version challenged Newton's theory of gravity, replacing it with the concept of curved space. Its mathematical expressions disclosed some of nature's most intimate secrets. One of its equations became the backbone of the atomic age.

Physicists have a special respect for relativity because, unlike other theories that fade away with time, relativity has become more relevant today than in its early years. Black holes, expanding universe, the big-bang theory—all borrow heavily from relativity. Experiments with giant atom smashers have verified its prediction that particles traveling at high speeds experience a growth of their mass. Atomic clocks that can measure time to a billionth of a second show that at high speeds time is indeed slowed down, as predicted by relativity. Recent experiments have placed relativity under all kinds of tests, but in each case it has emerged intact.

Relativity's chief architect was Albert Einstein, a brilliant scientist, a concerned citizen, and a warm human being. He was a quiet, good-for-nothing student who detested the regimented rote learning experiences of his youth. He sharpened his most powerful assets—concentration and mathematical ability—into the tools that he needed to look into the mysteries of nature. In his lifetime he was crowned as the prince of scientists and called the master-guru of science and the father of the atomic age.

The Speed of Light, Ether, and the Michelson-Morley Experiment

The heart of relativity is light. Light is the carrier of information that describes the world around us. Its inexplicable property of having the same speed regardless of the relative motion of its source and the observer was a key discovery that led to relativity. The debates earlier on were whether it travels with a finite or an infinite speed, and whether it was to be considered as a wave or as a particle. Both these questions played a major role in the preparation of the ground on which relativity was eventually built.

As early as 300 B.C., when Euclid was writing his book on geometry, Greek philosophers were using mathematics to study the reflection and refraction of light. Aristotle, like most other Greek philosophers, believed that light traveled instantaneously, with infinite speed. More than a thousand years later, the Arab scientists Avicenna (980–1037) and Alhazen, whose work in optics was extensive, reasoned that although its speed is very high it is nevertheless finite. Both, however, failed to support their proposition experimentally.

The first to propose an experiment to measure the speed of light was Galileo Galilei (1564–1642). In his book *Dialogues Concerning Two World Systems* (1638), Galileo describes a night experiment carried out by two people with lanterns on two hilltops. After some practice, one observer measures the time from the moment he uncovers his lantern until he sees the light from the observer on the other hill, who uncovers his light the moment he sees the light from the first lantern. In essence, the measured time is that needed for light to travel back and forth, twice the distance between the two hilltops. The velocity of light will be, of course, the distance divided by the time taken. The results were disappointing. In Galileo's words, "...I have tried the experiment only at a short distance, less than a mile, from which I have not been able to ascertain with certainty whether the appearance of the opposite light was instant or not, but if not instant, it is extraordinarily rapid."

Galileo's experiment failed because it used only a short distance and a crude timing method; a greater distance would have compensated for at least part of the error. The Danish astronomer Olaus Rømer (1644–1710) used such a great distance—the one between Jupiter and the earth—to demonstrate in 1676 that light indeed travels with a finite speed. Rømer noticed that the innermost satellite of Jupiter takes 42.5 hours to make its orbit, going through an eclipse every time it goes behind Jupiter. Rømer also noticed that these eclipses occurred later and later than their predicted time as the earth, orbiting the sun, went farther and farther away from Jupiter. The maximum delay was 1,320 seconds when the earth was the farthest away from Jupiter. At this position light from Jupiter must travel an extra distance, the diameter of the earth's orbit, estimated then to be 182 million miles.

So the time delay of 1,320 seconds was due to the extra distance of 182 million miles. Rømer used these numbers in the equation velocity = distance/time and obtained a value of 138,000 miles per second for the speed of light. Although crude, and short by 52,000 miles per second, Rømer's discovery was a significant event in the history of science because it shattered the long-held belief that light travels instantaneously with infinite speed.

But Rømer's astronomical explanation of the finite speed of light did not persuade everyone. The skeptics could be convinced only by a terrestrial experiment, one that they could put their hands on. In 1849—173 years later—Armand Fizeau (1819–96), a French physicist, designed an extremely simple experiment that measured the speed of light at 194,000 miles per second. He used a fast-rotating toothed wheel with a speed that could be slowly increased to chop a beam of light into pulses and send them to a mirror six miles away. In such an experiment, when the returning pulses arrive so that they hit only the teeth of the rotating wheel, then the view is complete darkness. But at some point, as the speed of the wheel is slowly increased, light pulses start flashing through the wheel, like a stroboscope. This means that returning pulses now encounter a slot and that the time taken for their round trip is equal to the time of rotating the wheel by one tooth. In Fizeau's experiment, this time was 1/18,000 of a second.

Thirteen years later Jean Foucault (1819–68), another young Frenchman, modified Fizeau's experiment by replacing the toothed wheel with a rotating mirror. His experiment required a light path of only sixty-six feet, rather than six miles. The rotating mirror turned the reflected light beam by a small angle, which corresponds to a tiny time interval. Foucault measured the speed of light at 185,000 miles per second. Ten years later Marie Cormu, an experimental physicist at the École Polytechnique in Paris, repeated Fizeau's experiment with an improved toothed wheel and obtained an even more accurate result—186,000 miles per second.

While the controversy over the speed of light was finally coming to an end, another one brewing for some years was slowly coming to a boil and was destined to provide the key to the theory of relativity. This was related to the question of the nature of light. Newton suggested that light is like a stream of particles. He called them corpuscles. Christiaan Huygens (1629–95) proposed, however, that light is propagated in wave forms, and some nineteenth-century discoveries were strong evidence for the wave theory. These were the interference experiments of Thomas Young (1773–1829) and the brilliant mathematical synthesis of Maxwell, in which light was explained as a combination of electrical and magnetic disturbances, propagated in the form of electromagnetic waves.

In accepting the wave theory of light, physicists had to deal with a thorny new issue—that of the carrier of the wave. Just as air is the medium for sound waves, scientists thought, a medium must be responsible for the propagation of light waves. Since light travels through a vacuum as well as through such matter as air, water, and glass, its carrier medium must be omnipresent. Physicists decided to call this medium ether, and they endowed it with some bizarre physical properties. It was invisible and without any weight.

Since the speed of a wave through a medium depends on its mechanical properties, ether was thought to have an elastic strength greater than that of steel.

Since ether was thought to explain the propagation of light and possibly that of electricity and magnetism in terms of a mechanical model, many scientists were happy that finally all the physical phenomena could be explained in terms of Newtonian physics. This was a glorious but brief moment in history, when the universe was described in Newtonian concepts. Two such concepts were the absoluteness of time and space. Newton had proposed that space was a physical reality, stationary and unchanged. He could not support this premise with scientific arguments, but he was just as happy to borrow evidence from theology. To Newton, space was not a geometrical abstraction but the manifestation of God's presence everywhere in the universe.

Although ether was thought to be a necessary medium for the propagation of light, its existence was very difficult to verify. For many it was only a hypothetical element that physicists created to help them as a temporary solution, just as phlogiston and the caloric were once thought to be immaterial elements responsible for fire and heat. Two American physicists, Albert Michelson (1852–1931) and Edward Morley (1838–1923), carried out a historic experiment in 1881 to verify the existence of ether. Their reasoning was very simple. An observer on the surface of the earth should be able to detect an "ether wind" as the earth, orbiting the sun at 20 miles per second, flew through a stationary ocean of ether. If beams of light were fired in all directions, their speeds should differ as they traveled with, across, or against the ether, just as the speed of a boat depends on whether it is flowing with or against the current.

Initially they used one of Michelson's early apparatuses, made of a beam splitter (a partially silvered glass that reflects part of the initial beam and transmits the rest) and two mirrors a few inches away and at right angles to each other. After a single beam is split into two parts, each travels at a right angle to the other and is then reflected back by one of the mirrors. When the two beams are brought together, they cause a phenomenon known as interference, in which a pattern of dark and bright fringes is formed. If the velocity of one of the beams were to change, then the fringe pattern would be expected to change also.

Michelson and Morley thought that any small rotation of the apparatus would change the velocities of the two beams (since their orientation to ether would be different) and therefore the fringe pattern inevitably had to change also. But this was not to be so. No matter how much the apparatus was rotated, there was no change in the fringe pattern. Michelson and Morley designed a more elaborate apparatus with many more mirrors, more accurate, and completely free of vibrations. After thousands of trials, the results were still the same. No matter how light traveled, with or against the ether, its velocity was always the same. It is just as if a spaceship traveling at, say, 100,000 miles per second toward or away from earth sends a beam of light toward us and we measure the speed of light always at 186,000 miles per second. Later experiments have indeed shown that no matter what the relative speed of the light source, the speed of light is measured to be the same by all observers.

The Michelson-Morley experiment confronted scientists with an embarrassing decision: they had to abandon either the theory of ether or, worse still, the more venerable proposition of Copernicus that the earth is in motion. No matter how preposterous the second alternative, some found it easier to accept than a world without ether. It is so ironic that Michelson and Morley initially pursued a noble task—to verify the existence of ether—yet ended up signing its death certificate. Little did they know that their results would provide the seeds of one of the greatest theories of physics, the theory of relativity.

The Special Theory of Relativity

The failure of the Michelson-Morley experiment to detect ether turned out later to be the greatest of all negative results in the history of science. The constancy of the speed of light, as suggested by their experiment, was to many scientists an enigma and an aberration of nature. But to Albert Einstein, a twenty-six-year-old patent-office clerk in Bern, Switzerland, it was a revelation of a universal law that was to become the pillar of his theory of relativity. In his historic 1905 paper "On the Electrodynamics of Moving Bodies," Einstein rejected the concept of ether and along with it the Newtonian concepts of absolute space and time. He proposed instead that observed mass, length, and time vary according to the relative motion between the observer and the system under observation.

The special theory of relativity is based on two fundamental principles. The first is that the speed of light is the same no matter how it is measured. The second is that all laws of nature are obeyed in an identical manner in all inertial systems; that is, those that are at a constant speed with respect to one another. As a consequence of these postulates, the picture of the universe through the window of a fast-moving rocket is distorted to a degree that depends on the speed of the rocket. At slow speeds the picture of the universe is no different from that viewed at rest, so relativity and Newtonian physics are in agreement. At high speeds, however, Newtonian physics fails in its predictions and descriptions of the universe. It is therefore inaccurate to claim that relativity displaced Newton's picture of the universe. It merely extended it to some very special situations.

Relativity is indeed an extraordinary theory with many "unreal" propositions. How can the mass and length of an object vary according to its speed? How and why should a clock in motion slow down? These are undoubtedly revolutionary ideas that run contrary to common sense. But Einstein warned us that common sense is nothing more than an assemblage of mental prejudices collected from everyday experiences since childhood. These experiences are only a tiny fraction of a rich and wide spectrum of universal events. One such prejudice is that no matter how or where we measure time, clocks will tick at the same rate. Relativity, however, states that the faster a clock moves the more slowly it ticks, something that can indeed be verified experimentally. Relativity also states that a moving rod shrinks more and more in the direction of motion as its speed increases. Also, the mass of a moving object grows with speed, again an "unreal" proposition, yet supported by substantial experimental evidence.

Relativity helps us see the fallacy of claiming any kind of absolute motion. In the absence of a universal stationary landmark, all motion becomes relative. For years the earth was thought to be stationary and was used as a universal landmark. Copernicus changed all that in 1543. We now know that the earth spins around its axis at a speed of 20 miles per second. It also orbits the sun at a speed of 1,000 miles per hour. But that is not all the motion of the earth. Our solar system moves with a speed of 13 miles per second within the local star system, which in turn moves within the galaxy at 200 miles per second. Our galaxy moves within a local group of galaxies at 100 miles per second.

So what is the overall speed of the earth? In which direction? With respect to what landmark? There is no such thing as absolute motion in the universe, only relative to another system.

Fizeau's experiment demonstrated that light is not transmitted instantaneously but has a finite speed. This was also the final blow to the long-held notion of a universal "now." We all know that the sky at night is a picture that is thousands and even millions of years old, because it takes the light from those distant stars eons to reach us. The picture of the night sky that is "now" to us is history to those stars. Some of the stars we see may not even exist anymore. Younger stars whose light has not reached us are not in the picture yet. So, in essence, past, present, and future are to some degree the separation of space as much as time.

The end of a universal now brought about the end of yet another time-related universal notion: that of universal simultaneity. Two events that appear as occurring simultaneously to one observer may be minutes, days, or even years apart to another observer. Einstein explained this using as an illustration a high-speed train hit by two lightning bolts—one at each end. To an observer on the ground, these may appear to have occurred simultaneously. But to a passenger in the middle of the train, the lightning bolt at the engine appears to be first. This is because the motion of the train shortens the distance that the light from the engine, carrying the message of the event, will have to make, whereas the light from the back of the train has to travel a longer distance. The faster the train, the longer the interval between the two events to the passenger.

Perhaps the most fascinating phenomenon in relativity is the slowing down of moving clocks, referred to as time dilation. To understand how this can happen, consider the following event. A flashlight inside a fast-moving rocket emits a light pulse at a right angle to the direction of the rocket's motion. The pulse is reflected by a mirror on the opposite wall and returns to its source. This is the event. Its duration is measured by two observers. One is an astronaut inside the rocket, and the other is on earth watching the rocket passing by in front of him. What the astronaut saw was a light pulse that went up and down, traveling the width of the rocket back and forth. The duration of the event for him was (according to the formula distance = velocity × time) equal to the distance (twice the width of the rocket) divided by the speed of light (3×10^8 m/s).

But the observer on earth sees a different picture. The pulse did not just go up and down; it also traveled a horizontal distance from left to right because of the motion of the rocket. For this observer the path will look like an inverted V made by the combination of these motions. This is a longer distance than the one measured by the astronaut. Since the velocity of light is the same for all observers, the observer on earth will measure the duration of the event (distance over velocity of light) as much longer than the astronaut will. Of course, the faster the rocket travels the longer the inverted V and therefore the longer the event will appear to the stationary observer.

Now if the up-and-down journey of the light pulse is a unit of time for some unusual light clock, then what may appear as a normal time unit to the astronaut will appear to last longer to an outside observer, as if the clock were ticking slowly. Time dilation has nothing to do with the structure of the clock. All moving clocks, whether mechanical or biological (like the heartbeat or the life cycle of a cell), are not only seen by outside observers to slow down but really do, as various experiments to be described later show. So traveling at speeds near that of light may be one way of extending life, as compared to that of a stationary observer.

Such fascinating ideas have been the theme of many science-fiction stories. In the movie *Planet of the Apes,* for example, an astronaut returned to earth to find out that while he had aged by only a few years, many more years had passed on earth. Another popular example used to illustrate biological time dilation is the one in which an astronaut after only a few years of high-speed space travel returns to find his twin brother an old man.

Time dilation is a hard phenomenon to accept because of our deeply rooted reverence for absolute time. But just as color is very much related to relative motion (the Doppler effect: the faster a light source comes toward us, the bluer its color; the faster it travels away from us, the redder its color), time is also a form of perception that depends on relative motion. Just as space is measured by a sequence of landmarks, time is measured by a sequence of events. But since events take place somewhere, they have spatial coordinates. So time and space are intimately related. What we call an hour, for example, is connected to a measurement of space, a 15-degree arc in the daily rotation of the earth. We call a year the earth's completion of a journey through a specified space, an elliptical path around the sun. But this earthly time makes no sense on another planet. The motion of the planet Mercury, for example, is such that the time for one revolution around the sun is equal to 88 earth days and one rotation on its axis takes 59 earth days. Thus, a Mercury day is 59 earth days and a Mercury year is only 88 earth days.

In yet another bizarre relativistic effect, a rod traveling at, say, 90 percent of the speed of light will "shrink" to about half its original length, the contraction taking place only in the direction of motion. Such a contraction, which depends on the speed of the rod, is neither real nor an optical illusion. Relativity simply states that the motion distorts space—in this case, the length of the rod—in a similar way that it distorts time. Someone riding with the rod will notice no change in its length.

To understand how traveling rods shrink, consider the following explanation. Suppose that the rod has two identical clocks, one at each end. An outside observer notes the time that each clock shows as it passes in front of him at the velocity V. Let Δt be the time difference shown by the clocks as they go by the observer. It takes time (Δt) for the rod to fly by the observer. Its length therefore will be $V \times \Delta t$ (distance = velocity × time). Since the faster the rod, the smaller the value of Δt (clocks slow down), it follows that the observed length shrinks more and more as the rod travels faster and faster.

Relativity seems foreign to our everyday common experiences because its phenomena are extremely difficult to detect. It takes six years even for a trained physicist to feel comfortable with it, not because of its mathematical complexity but because of its strange ramifications. To top it all, this Alice in Wonderland world of high speeds exhibits the property of reciprocity. An observer on earth measures the mass of a fast-moving spaceship as having increased, its length as having contracted in the direction of motion, and time aboard the ship as running slower. But for an astronaut inside the spaceship none of these takes place. On the contrary, when he looks outside he finds that mass, time, and length on earth have changed in exactly the same way. Of the two observers, which one has the truth? According to relativity, both.

But how can two different world views be both right? Observing the decay of particles called muons can explain this paradox. When they are formed in the laboratory, these subatomic particles have a lifetime of only two microseconds (a microsecond is a millionth of a second). But when they are created in the upper atmosphere by cosmic rays they have speeds close to that of light and therefore have an "extended" lifetime of sixty microseconds. This enables them to travel through the atmosphere and reach the surface of the earth. To an observer on earth, the traveling muon constitutes a clock that slowed down owing to its high speed. But from the point of view of the muon its lifetime is still two microseconds. What it "sees" as different is the thickness of the atmosphere, which appears to have contracted. The muon therefore sees nothing strange about reaching the surface of the earth. Similar muons when accelerated to high speeds in particle accelerators have verified Einstein's formula of time dilation to an accuracy of 0.2 percent.

Subatomic particles can be made to travel at high speeds, which makes them ideal for demonstrating relativistic changes. Our bodies, which by comparison can attain only low speeds, find it difficult to witness at first hand changes such as the slowing down of our own "biological clocks." For example, an astronaut who orbited the earth for 80 days at 8 kilometers per second (which is only 0.003 percent of the speed of light) has stayed 0.002 of a second younger than her twin on earth. Even these small changes can now be measured with atomic clocks that are accurate to a billionth of a second. In 1972 a scientist from the U.S. Naval Observatory took such a clock with him on an airplane trip around the world. After the trip, it was compared to another one left on the ground. The traveling clock had slowed down by exactly the amount predicted by relativity.

The most convincing proof of relativity comes from radioactivity and nuclear energy. $E = mc^2$, the conversion of mass into energy, is relativity's most fascinating formula. It is derived from a rather unusual behavior of matter, in which the mass of an object increases with its velocity. This suggests that mass can actually be created out of energy. So a hot iron is heavier than a cold one, but only by less than a billionth of a gram. When the *Apollo* spacecraft was hurtling toward the moon at thousands of miles per hour, its mass was heavier by a few thousandths of a gram. When small particles, such as electrons, are made to travel near the speed of light at the Stanford accelerator, their mass increases as much as forty thousand times. As the mass grows, so does its resistance to acceleration, a quality known as inertia. It therefore becomes more and more difficult for an object to increase its speed further. It is impossible for a material object ever to reach the speed of light.

At least in theory, phenomena beyond the speed of light belong in a "forbidden" zone that nature prevents us from experiencing. The speed of light is one of nature's boundaries that pose a mystery to science and an intriguing topic to philosophy and metaphysics. At least another such boundary exists in low temperatures. It is spelled out by the third law of thermodynamics, which states that it is impossible to reach absolute zero (−273°C). As in relativity, where space, time, and matter behave in an unconventional manner at high speeds, matter exhibits some peculiar properties at very low temperatures. It is as if these two boundaries, and maybe more still to be discovered, form a mysterious confinement that separates our world from other strange neighboring worlds.

The General Theory of Relativity

The publication of the special theory of relativity in 1905 did not make Einstein an instant celebrity. He continued to work for four more years as a patent clerk, supporting a wife and two children on a mediocre salary, before he was finally offered an academic post. Slowly but steadily Einstein was gaining respect and recognition from renowned physicists everywhere. Yet he knew that his special theory of relativity was both limited and incomplete. It took him ten years, from 1905 to 1915, to produce the general theory of relativity, a more encompassing theory that has been described as the greatest triumph ever achieved by the disciplined imagination.

Those were years of mixed blessings in Einstein's life. His marriage was failing. He had been offered the directorship of the Kaiser Wilhelm Institute in Berlin, one of the most prestigious science research centers in Europe, and he signed a manifesto denouncing World War I, declaring his pacifism publicly for the first time. But above all he was rolling back the frontiers of science by developing a new theory of the universe. He had to struggle through unfamiliar territory, using the new mathematics of Riemannian geometry and tensor calculus before he finally produced his masterwork, the general theory of relativity.

In a letter to a friend, while he was still working on his new theory, Einstein wrote, "In all my life I have never before labored so hard.... Compared with this problem the special theory of relativity is child's play."[12]

The "special" in the special theory of relativity refers to only limited cases of motion—those with constant speed. The special theory did not take into account accelerated motion or motion in a variable gravitational field. To Einstein this was not only incomplete but aesthetically unsatisfying. His instincts for beauty, unity, and universality in nature convinced him that there must be a more general theory that could explain all cases and all phenomena.

The starting point in his thoughts was the phenomenon of artificial gravity, which we all experience sometimes and take so much for granted. When an elevator suddenly accelerates upward, we feel momentarily heavier. In essence, gravity can be made artificially by an accelerated system. The connection between acceleration and gravity is described in the principle of equivalence, which is a cornerstone on which the general theory of relativity is built.

The principle of equivalence states that observations or experiments made in an accelerated system are indistinguishable from those made in a gravitational field. In other words, the effects of acceleration and gravity are identical. When an astronaut floating under zero gravity in a spaceship suddenly "falls" to the floor, he will not be able to tell whether that was caused by a sudden acceleration of his spaceship or the spaceship's landing on a planet. Einstein concluded that this principle holds true not only for mechanical phenomena but also for all natural phenomena—optical, electrical, and even biological.

Let's apply the principle of equivalence inside a large rocket, which at first is resting on earth. Suppose that we throw a ball horizontally to the other wall. The earth's gravity will bend the flight path of the ball as it does that of a ball rolling off a table. According to the principle of equivalence, the ball would have the same flight path if the rocket were traveling in outer space with an acceleration equal to that caused by the earth's gravity (9.8 meters per second^2). If we now replace the ball with a beam of light, we must detect a small but measurable bending here too—because light, having energy, exhibits mass-related characteristics, according to $E = mc^2$. The extraordinary conclusion from the principle of equivalence is that if light bends inside an accelerated rocket, it must also bend in a gravitational field.

It is very difficult to detect the bending of light in a small gravitational field such as earth's. If we fire a laser beam at the earth's horizon, it will drop a mere ¾ inch over a 4,000-mile straight path. Since the sun generates a much stronger gravitational field, light passing by it is expected to bend more. As long ago as 1915, Einstein predicted that starlight passing close to the sun would be deflected by 1.75 seconds of arc—a very small angle indeed, but large enough to be measured on earth.

Since stars are invisible during the day, their light deflection can be observed only during a total solar eclipse, and the English scientist Arthur Eddington (1882–1944), fascinated by Einstein's prediction, organized an expedition to observe the 1919 solar eclipse. Photographs of the "darkened" sun taken during those precious moments of the eclipse showed that the position of nearby stars was slightly displaced by the bending of their light, confirming Einstein's prediction. Einstein was a very happy man when he heard the news. In a telegram to his mother he said, "Joyful news today. H. A. Lorentz has telegraphed me that the English expedition has really proved the deflection of light by the sun."[13] This was the first real challenge to Newton's theory of gravitation.

It was the prestigious Royal Society of London, of which, ironically, Newton had been president some two hundred years earlier, that formally announced to the world Eddington's results verifying Einstein's theory of relativity. Einstein was instantly hailed as a genius and the most famous living scientist. But even without Eddington's data Einstein had known that his theory was correct. After all, it explained why the orbit of Mercury precesses by forty-three seconds of arc every century, something that had puzzled astronomers for hundreds of years (precession is the slow, steady motion in which the axis of a spinning object, like a planet, describes a circular cone about the vertical).

The story goes that when a student asked him what would have happened if the solar expedition had not confirmed his predictions, Einstein replied, "Then I would have been sorry for the dear Lord; the theory is correct."[14]

Since then more accurate measurements have been made using other regions of the electromagnetic spectrum than visible light. Radio-emitting stars known as quasars have been used as markers. As their radio signals pass near the sun they bend to within 1 percent accuracy of Einstein's theory.

If light has mass characteristics, as shown by its bending by the sun, when it falls into a strong gravitational field its velocity should increase, just like that of a falling rock. Likewise, when it leaves a gravitational field its velocity should decrease again, just like the velocity of a rock thrown straight up. We know, of course, that nature forbids that. No matter what, the velocity of light in a vacuum remains constant. (When light travels through other media such as water or glass, its velocity decreases.) If the velocity of light is unaffected by gravity, then what other changes can take place when light falls into or escapes from a strong gravitational field?

Einstein predicted that the change would be one of frequency or color. As light escapes a gravitational field, its color becomes slightly redder. This phenomenon is known as gravitational red shift (not to be confused with the red shift of the Doppler effect; the first is caused by gravity, the second by the relative motion between the light source and the observer). Einstein explained gravitational red shift using the example of an elevator accelerating upward. Consider a rhythmic signal being sent from the floor to the ceiling. An observer at the ceiling will receive these signals at a slower rate because he is accelerated constantly away from the floor. In other words, if five signals are sent from the floor each second, the upward acceleration may allow only three to reach the ceiling each second. If the signals were light, then the lower the detected frequency (vibrations per second), the redder the light. By the principle of equivalence, the situation would be identical if the elevator were not accelerating but were stationary on earth. Since gravity decreases with height, the ceiling will be in a slightly smaller gravitational field than the floor. So light emitted at the floor should be slightly redder when it arrives at the ceiling.

Such gravitational red shifts are extremely small and unnoticed by our eyes. Delicate instruments can detect and measure by exactly how much the color or frequency of light changes. Einstein predicted that the color of light emitted by atoms on the surface of the sun, where gravity is very strong, should show a tiny red shift compared to the light from similar atoms in the laboratory. Constant turbulence on the sun's surface made such testing difficult, but later improved methods verified Einstein's prediction once again.

A more convincing demonstration of gravitational red shift was carried out at Harvard University in the 1960s, a few years after Einstein's death. The demonstration made use of a newly discovered phenomenon, the Mössbauer effect. Rudolf Mössbauer (1929–), a German physicist, discovered in the 1950s that some atoms that emit gamma rays can reabsorb their own radiation provided that it has the same frequency with which it was emitted. In the Harvard experiment one such gamma source was at the bottom and another at the top of a twenty-three-meter tower that was evacuated to eliminate the effects of air. It was found that the source at the top would not absorb the gamma rays emitted by the one at the bottom of the tower. But when the sources were brought side by side, absorption could take place. It was fascinating that only twenty-three meters of gravitational change were enough to demonstrate convincingly Einstein's gravitational red shift. The gamma ray rising through twenty-three meters changes its frequency only slightly, but enough to be unrecognized and rejected by the source at the top of the tower.

The electron vibrations in atoms that produce light are like the steady ticktock oscillations of a mechanical clock, only at a much higher frequency. So atoms are really miniature, superaccurate clocks. Now if gravitational red shift affects atomic clocks, it must also affect any other clock, mechanical or biological. So clocks in a strong gravitational field, as on the surface of the earth, must slow down compared to those in a weaker gravitational field, such as higher up from the ground. When two atomic clocks are placed one at the top of a skyscraper and the other at the bottom, they show different time, though the one at the bottom slows down only by billionths of a second. The same is true, of course, for biological clocks. A secretary who works on the ground floor will be slightly younger than her twin sister who works on the top floor, but only by a few millionths of a second in a lifetime. Clocks in the vicinity of ultra-strong gravitational fields, like those around a black hole, slow time to come almost to a standstill.

This is indeed one of the most fundamental differences between Einstein's and Newton's views of gravity. For Einstein, because gravity affects light, it also affects time. Since the strength of gravity decreases as the square of the distance from its source, large objects like the sun or even a black hole are surrounded by shells of different time zones.

Gravity's influence on time and light were seen by Einstein not as some isolated aberrations of nature but rather as manifestations of a grand design that encompasses all the phenomena of relativity. From 1911 to 1916, Einstein searched for a model that could explain gravity and its effect on light and time in a simple and comprehensive manner. He finally developed a four-dimensional continuum of space and time, where space was "curved" by gravity. Einstein described a universe in which gravity is no longer a force but geometrical distortions in a multidimensional space-time continuum.

Although the mathematics required were already available, their application to the complex world of relativity was an intellectual challenge even for a genius. Einstein, who considered himself a physicist rather than as a mathematician, was fortunate to have the help and support of his dear friend Marcell Grossmann (1878–1936), a brilliant mathematician. They were both professors at the Zurich Polytechnic, where Einstein taught for some time. Years earlier, when they were students there (Einstein had failed his first entrance exams), Grossmann coached Einstein in mathematics. Grossmann was the mathematical expert of the team, knowledgeable in non-Euclidean geometry and tensor calculus, essential mathematical tools for relativity. Einstein provided the physical insight and direction that were often based on aesthetic rather then empirical arguments. At times when they were on false tracks, it was Einstein with his perception of nature's simplicity, symmetry, and beauty who reoriented them in the right direction.

The non-Euclidean geometry that Grossmann and Einstein used was a well-established branch of mathematics. Whereas Euclidean geometry deals with flat surfaces, non-Euclidean geometry deals with curved surfaces. The rules in each geometry are very different. For Euclidean geometry, they are also very rigid. For example, all angles of a triangle add up to 180 degrees, the shortest distance between two points is a straight line, and the ratio of the circumference to the diameter for any circle is 3.14 (pi, π). If the same triangle is drawn on a curved surface, such as that of a sphere, it has entirely different properties. For example, the sum of its angles is greater than 180 degrees, and changes according to the curvature. The distance between two points on the triangle is no longer a straight line but a curve. These lines that take the contour of the surface are called geodesic lines—geodesics, for short.

In Einstein's universe, gravity, space, and time are interrelated, forming the four-dimensional space-time continuum. Our brain, evolved in a three-dimensional world, finds this incomprehensible. Any attempt to visualize it therefore is futile. All we can do is understand some of the properties of this new model of our universe. The first thing to appreciate is the role of time as a dimension. On a map we need only two coordinates to locate a point. On a three-dimensional map, we need a third dimension: height. With time now as a fourth dimension, we are never in the same coordinates in the space-time continuum even if we are stationary. As time flies, one of our coordinates and therefore our position changes. Consider as a very simple example a caterpillar going round and round on the rim of a coffee cup. To our view it goes in a circle. In the space-time continuum its motion is described as a helix, with time the vertical axis.

In Newton's universe a massive object like our sun appears as matter generating a force of gravity that can affect other objects. The gravitational force gets weaker as the distance from the sun increases. But according to Einstein, the sun distorts the space-time continuum into a "curved" surface, or warp. The curvature is large near the object, where gravity is strong. Objects no longer experience a force but follow the path of the curved surfaces, or geodesics, which deflect their path. According to author Nigel Calder in *Einstein's Universe*: "Gravity rules the universe, but according to Einstein it exerts its authority not by force but by persuasion—the ineluctable persuasiveness achieved by laying down convenient tracks through space and time."

One of the remaining mysteries of relativity is how gravity is transmitted through space and matter. Einstein proposed that gravity must have some common features with light. For example, it has been suggested that its speed of propagation is the same as that of light. Since light exhibits both wave and particle characteristics, the same is expected of gravity. The "particles" thought to be responsible for the transmission of gravity are called gravitons. But their nature and properties remain a mystery, drawing little attention from theoretical and experimental physicists.

The wave nature of gravity, on the other hand, has attracted a lot of attention, and a number of experiments are already under way to detect gravity waves. In theory, every time an object makes a sudden shift in position it generates gravity waves, "ripples" through the space-time continuum. Even waving good-bye produces tiny gravity waves, which in theory can travel over a long distance. Massive gravity waves are generated when a large star "dies" in a spectacular supernova explosion. Like earthquakes through the earth's crust, gravity waves should distort space as they pass through it.

In a typical experiment set up to detect a galactic gravity wave, a large aluminum cylinder floating in a vacuum is used as a sensor. When gravity waves pass through it, they are expected to cause a minute distortion that signals their arrival. Although current experiments have not as yet been able to detect gravity waves, improvements in the sensitivity of the sensors are expected to make such detection possible in a few years.

Relativity continues to be a hard theory to accept because its effects are beyond our everyday experiences. Yet its validity has been demonstrated repeatedly, particularly since Einstein's death. On a cosmological scale relativity can explain some of the bizarre behavior of black holes, collapsed stars with such a strong gravitational field that even light cannot escape. Relativity provides us also with some reasonable theories about the birth and expansion of our universe. But, like any other scientific theory, even relativity one day will have to be revised. Theoretical and experimental physicists continue subjecting it to rigorous tests.

One such experiment currently under preparation at Stanford University will check gravitomagnetism, a phenomenon predicted by Einstein, caused by the rotation of the earth. Researchers plan to use supersensitive gyroscopes spinning in space at ten thousand revolutions per minute. A small shift of the axis of the spin would be in accordance with Einstein's prediction. Gravity probe B, as the experiment is called, is planned to have a prototype gyroscope tested in space through NASA's space shuttle program by 1990. The year-long experiment is scheduled to be conducted sometime in the early 1990s.

Relativity offers a very unorthodox picture of the universe. Its revolutionary concepts have given birth to fascinating relationships between what were once thought to be completely unrelated concepts. Energy and mass are manifestations of the same thing. Time

Albert Einstein with J. Robert Oppenheimer, the director of the Manhattan Project. (Courtesy American Institute of Physics, Niels Bohr Library)

depends on the speed of the clock and on gravity. Space is defined by gravitational fields, which are related to matter, which can be regarded as a concentrated form of energy. Matter, energy, space, time, and gravity are like musical instruments that play the symphony of the universe, with light as their conductor.

The influence of relativity went beyond physics into such areas as the arts, philosophy, and even politics. Philosophers, for example, used it either to support or to denounce materialism. For some, its strong emphasis on mathematics and the lack of mechanical models implied a rejection of the mechanistic view of the universe and with it the materialistic philosophy. Others saw relativity as implying an end to absolutism, fundamentalism, and all ethical codes. Everything was thought to be considered relative. These ethical skeptics were often supporters of a materialistic philosophy. So Einstein's theories were used equally as arguments for and against materialism, often surfacing even in the propaganda machines of communism and fascism.

It was a dark day in the history of science when prominent German scientists, joining the Nazi frenzy, denounced relativity as a Jewish, anti-German theory. Physicist Philipp Lenard, a Nobel Prize winner, wrote, "The relativity theory of the pureblooded Jew Albert Einstein when faced with reality has no leg to stand on.... In contrast to the equally intractable and solicitous desire for truth of the Aryan scientist, the Jew lacks to a striking degree any comprehension of truth...."[15] In the 1938 edition of the *Journal for General Science*, founded to promote the Nazi philosophy of science, it was written, "Einstein's theory of relativity offers us the clearest example of dogmatic Jewish type of theory. It is headed by a dogma, the principle of the constancy of the velocity of light."[16]

Relativity was even given a religious flavor. In Nazi propaganda articles, "Jewish physics" was seen as "Talmudic thinking,"[17] an attempt to fulfill the Torah of Judaism. Roman Catholic theologians argued whether relativity was materialistic or idealistic. Some even saw the four-dimensional concept as a vision of the afterlife. A director of a theology department at a British college wrote in a 1939 edition of the *Hibbert Journal* (a quarterly review of religion, theology, and philosophy published in London), "While we are limited to three-dimensional understanding, it is mortal life. Where we perceive it in four dimensions it is eternal life."

Historians of science suggest that great breakthroughs can be made only if the intellectual climate is right; under these conditions, the birth of a new theory is almost inevitable. It is probable that if Galileo, Newton, or Faraday had never lived, others would have produced their theories. If not Einstein, someone else eventually would have solved the mystery of ether. The general theory of relativity, however, is unique, and it belongs exclusively to Einstein. It took his brilliant imagination and bold intellectual courage to synthesize space, gravity, and time. Without Einstein, the general theory of relativity might still be unknown today.

When he was five years old Einstein was stirred by the sight of a compass. The mysterious force that turned the magnetic needle fascinated him. Although his teachers thought of him as good-for-nothing and a lazy dog, Einstein developed a passion for geometry and a deep curiosity, which he expressed with simple questions. At age sixteen, he asked what the world would look like if he rode on a beam of light. After ten years, his theory of relativity provided the answers. Along with his talents, Einstein developed a rich imagination and a healthy dose of skepticism. His work in the patent office helped him sharpen his analytical skills. Reviewing applications for patents, he had to go to the heart of the problem, searching for fallacies, inadequacies, and weaknesses. These were the same tools with which he challenged the old physics.

But behind this genius there was another Einstein, a kind and warm human being full of compassion for all who suffered, with a deep commitment to human values. Despite his worldwide fame, he lived humbly and simply. He had a modest house, and his luxuries were smoking cigars, playing his violin, and sailing his boat. His passion for simplicity often took a twist of eccentricity, exemplified by his refusing to wear socks, letting his hair grow, and showing little interest in his wardrobe. His philosophy of life could have been extracted from a religious text. He said, "I have never looked upon ease and happiness as ends in themselves—this ethical basis I call the idea of a pigsty. The ideals that have lighted my way, and time after time have given me new courage to face life cheerfully, have been Kindness, Beauty and Truth.... The trite objects of human efforts—possessions, outward success, luxury—have always seemed to me contemptible."[18]

Beneath that sweet, soft-spoken personality there was a rebel Einstein whose unpopular crusades drew fire from many quarters. He had a lifelong distaste for the military system—"the plague-spot of civilization," as he called it—and any form of authoritarian tyranny, whether fascism or communism. He was himself the victim of such a system. When he left Germany for the United States in 1933, the Nazis put a price on his head, and his house was confiscated. Throughout his life, he was a devout pacifist who advocated disarmament and international cooperation. In his later years, he proposed a world government. He strongly supported the establishment of the state of Israel, and was invited to become its second president.

Although he was born to a Jewish family, Einstein never really expressed a keen interest in any particular religion. He rejected the idea of an anthropomorphic God "who rewards and punishes his creatures, or has a will of the kind that we experience in ourselves."[19] Instead, Einstein chose to believe in "Spinoza's God who reveals himself in the harmony of all beings, not one who concerns himself with the fate and actions of men."[20] Einstein's God was really nature itself, with all its mysteries, laws, and marvels. Doing science was almost a religious mission. "I maintain that the cosmic religious feeling is the strongest and noblest motive for scientific research,"[21] he said.

As for the conflict between science and religion, Einstein blamed it on ignorance and arrogance. He viewed them not as contradicting each other but reinforcing and complementing each other, each satisfying a different need of the human condition. "The situation," he said, "may be expressed by an image. Science without religion is lame; religion without science is blind."[22]

Gravely concerned about the escalation of domestic and world problems, Einstein considered education one of our surest hopes for bringing about peace and prosperity. He believed that along with skills and specialized knowledge education should develop harmonious personalities with an understanding of and feeling for values. To Einstein the most critical condition for such an education was the encouragement of curiosity in an atmosphere of intellectual freedom. Once he wrote, "It is almost a miracle that modern teaching methods have not yet entirely strangled the holy curiosity of inquiry; for what this delicate little plant needs more than anything, besides stimulation, is freedom."[23]

Although many countries have claimed him as their citizen and hundreds of people have called themselves his friend, Einstein was really a lonely and detached figure. Undoubtedly his scientific legacy will endure for many years to come; it is unfortunate that his human and philosophical legacy continues to be ignored by the many and appreciated only by the few.

Chapter Seven

The Cosmos

To our ancestors the sky was a place of mystery and power. This vast space that contained the stars, the moon, and the sun and gave birth to wind, rain, and lightning was seen as the habitat of gods. Early astronomy, like many other sciences, had a strong association with religion.

Today we look at the sky with an equally strong sense of fascination and curiosity, and despite the many discoveries, we admit that all we know about our universe is only a tiny part of the overall picture. Yet every new discovery, no matter how small, develops an inner satisfaction and fuels our curiosity in a way that transforms our passionate search for answers almost into a religious mission. As we scan the sky at night, we realize that we are looking not only into the deepest corners of our universe but also back in time, for the light that reaches earth now may have left its source eons ago. We have developed a sense of kinship with the rest of the universe, knowing that we share the same material elements, are governed by the same physical laws, and most likely evolved from the same primeval source. We hope to develop more kinships, particularly in finding out about life beyond our planet.

During the last few decades we achieved great strides in our cosmic mission. We left our footprints on the moon and sent a manufactured object, *Pioneer 10*, beyond our solar system. Powerful instruments have answered many of our questions, but at the same time revealed a universe that is mysterious, even bizarre. New questions about peculiar objects such as pulsars, quasars, and black holes challenge us. The universe may be playing a game with us, in that the moment one of our questions is answered, hundreds of others are born. Its message may be that our quest may never have an end. Our questions multiply and become more complex.

Two questions have remained prominent throughout history, and no doubt will be asked by our sophisticated descendants: Where did this universe come from, and where is it going? These perpetual questions produced fascinating cosmological theories and myths and at times were blended into religious and philosophical musings on the purpose of the universe.

The Birth and Death of Stars

According to Aristotle, earth, the center of the universe, was made of the "corruptible" elements of earth, water, air, and fire; the heavens were made of pure and incorruptible ether. In addition, differences in the physical laws governed each domain. Aristotle's cosmos was combined with the later theory of Ptolemy (second century) on the geometrical paths of the heavenly bodies, and remained the dominant theory of the universe for almost two thousand years. Religious elements were added to it later to make it the official Christian model, in which God, angels, paradise, and hell had their geographical place and function. But this model faced major challenges in the sixteenth century. When Nicholas Copernicus (1473–1543) published his famous book *De Revolutionibus* (1543), the sun replaced earth as the center of the universe. Furthermore, the observations of comets and new stars were a severe blow to the incorruptibility of the heavens. When Galileo turned his telescope on the moon in 1610, he observed not the smooth surface of ether he expected, but a rugged and "decaying" landscape. That was the end of the purity and incorruptibility of the universe.

Today we know that stars are born, live their lives, and then die. They all start as huge gaseous clouds that contain hydrogen and cosmic dust. The mass of the cloud determines the amount of the inward-pulling force of gravity, which in turn initiates the birth, controls the life, and finally determines the type of death of each star. The massive cloud starts to collapse slowly under its own gravity. As it shrinks, its gravity grows stronger and the rate of collapse accelerates, causing the temperature and density of its center to increase. If the mass is large enough, the temperature reached will be in the millions of degrees, high enough to start thermonuclear reactions. Under these conditions hydrogen nuclei fuse together to form helium, a process that converts mass into energy according to $E = mc^2$. Since c^2 is such a large number (90,000,000,000,000,000), a small amount of mass becomes a very large amount of energy. The enormous thermonuclear energy produced convects outward, generating pressure that pushes against the force of gravity. The star continues to collapse until the crushing force of gravity is balanced by the exploding force of rising thermonuclear energy. The star, now in a stable state, will live most of its life under these conditions, constantly radiating energy.

Our sun is in such a state right now. Although small in size and going through its middle age at 5 billion years, it stands as the colossus of our solar system. Its powerful gravitational force keeps the planets in stable orbits, and its showers of energy and light make life possible for us on earth. Early humans deified, worshipped, and even sacrificed to it. The early Britons built Stonehenge as its temple, whereas for the early Greeks it was Apollo riding on chariots across the sky. The Aztecs sacrificed thousands, raising the freshly cut human heart toward the sun. For the Assyro-Babylonians it was their god, which came down to hand over to King Hammurabi one of the greatest of the ancient law codes. But even in modern times the sun was seen almost as a deity. Even Copernicus, who placed the sun at the center of our solar system, described it as a king sitting in majesty on a throne in the center of the universe.

The life of stars can be mapped on the Hertzprung-Russell diagram, bearing the names of its creators. The brightness and surface temperature of a star are the two important quantities used to describe its life in this diagram. Astronomers often referred to the brightness as magnitude and the surface temperature is often called color, since the intensity of the colors of the emitted spectrum depends on temperature. Most stars, when placed in this diagram, fall in on a diagonal band known as the main sequence, where a star spends most of its life. The end of a star begins when the hydrogen fuel in the core is diminished. The outward-pushing force, generated by the thermonuclear energy produced in the core, is reduced and can no longer counterbalance the inward gravitational force. The star begins to collapse, causing the temperature of its center to increase.

From here on the star-death sequence that follows depends on the mass of the star. If it is one to four times that of our sun, then the temperatures developed will be high enough to allow hydrogen in the outer layers of the core to fuse into helium. The extra energy produced forces the outer layers of the star to swell and redden. The star is now in a state known as a red giant. Our sun will reach that stage in about five billion years. As a red giant it will expand a hundred times in diameter, and its brightness will increase a thousand times. It will engulf the inner planets—Mercury, Venus, Earth, and Mars—and continue to burn for another hundred million years. In the meantime, the core will continue to collapse, developing such high temperatures that the helium fuses into carbon. This sudden production of extra nuclear energy, known as helium flash, will push the outer layers even farther away. These will continue to expand and eventually will separate from the collapsed center, giving the appearance of dust rings—nebulae. After a continuous expansion of fifty thousand years, the gas nebulae will become transparent and disperse into space.

The collapsing core, now known as a white dwarf, will continue to convert helium into carbon by nuclear fusion. Slowly the fuel will run out, and the sun will cool to its final stage, a black dwarf. It will become a star corpse—so dense that a teaspoon of its material may have a mass of a few thousand kilograms. This high-density carbon body may well be a giant nonglittering diamond in the sky.

A star whose mass is more than five times that of the sun ends its life more spectacularly. Since its large collapsing body generates ultra-high temperatures at the center, heavier elements such as carbon and iron are produced by nuclear fusion. Energy forces the outer layers to grow enormously, and the star becomes a red supergiant. But the rate of the core collapse is so fast that it allows no time for the energy to convect outward and radiate into space. Instead, the energy causes a spectacular explosion—a supernova, burning brighter than a billion suns. Such an explosion was first recorded on July 4, 1054, by Chinese astronomers. They reported a brilliant new star that shone more brightly than Venus and was visible even in daylight for several weeks. That place in the sky, now a huge thin cloud of dust, is known as the Crab Nebula.

The leftover core of a supernova continues to contract, becoming a star composed of neutrons. Basic laws of physics tell us that if a rotating body contracts, its speed increases, like ice skaters when they draw in their arms. So the neutron star spins faster, carrying along with it a strong magnetic field that produces strong radio signals as it sweeps through the electrically charged particles surrounding it. The star is known as a pulsar. When a pulsar was first discovered, in 1967, it caused great excitement, for at first it was believed that contact had been made with an extraterrestrial civilization. Pulsars still remain the mystery lighthouse beacons of the sky, sending out intense radio signals with clocklike precision, some as slow as one every four seconds and others as rapid as thirty per second.

Supernovas are thought today to be the factories of most of the heavy elements scattered throughout the universe. Elements heavier than carbon have been fused out of simpler elements under conditions of high temperature and density, found in supernova explosions. If this theory is correct, then newly formed stars must contain a higher percentage of heavy elements than old stars. Recent observations confirm this.

Practically everything that we use is made of atoms that are the fossil remains of a supernova explosion billions of years ago. Even parts of our body, such as our teeth and bones, the gold that we wear, and the car that we drive are all made of atoms that a supernova "cooked" out of hydrogen and helium in a process that alchemists would have envied. If we stop and think carefully for a moment, we realize that indeed we are all made out of star stuff.

If the mass of the star is thirty to fifty times as great as our sun's, its core will collapse under the enormous gravitational force. Not even the energy generated can slow down the crushing of the core. The laws that govern the behavior of matter and energy under these conditions are still a mystery. Using the standard laws of physics, we deduce that the collapse can go on indefinitely, reducing the star to a point—or, as physicists call it, a singularity. Although infinitesimal in size, it has the mass of a giant star. Its huge gravitational field attracts anything that falls within its influence, whether matter or light. Light can indeed be bent by a large gravitational field, as proposed by Einstein's general theory of relativity and verified by Eddington's observations in 1919. Since light cannot escape from this black abyss and light and other matter falls toward it as in a giant vortex, physicists call it a black hole. We know of its existence through the observable effects that it has on its neighborhood. For instance, a nearby star may develop a tail, losing mass into this cosmic hole. Matter accelerating toward it can produce X-rays, which, if produced far enough away from the black hole, can escape, carrying the testimony of its existence. At other times the black hole acts as a giant lens, bending starlight that passes through its gravitational field. Any such light received on earth gives a mirage image of its source, since eyes can only extrapolate light back in straight lines.

Black holes are scattered across the universe, and they come in a variety of sizes. The latest cosmological theories suggest that myriads of them were born at the birth of the universe. Some are as small as a proton but as heavy as a mountain; others are as big as Mars, with the mass of 100 million suns.

In addition to astronomers and astrophysicists, theoretical nuclear physicists believe that many of their questions are locked inside black holes. Working toward a unified field theory—one that explains all the major forces in the universe (gravity, electromagnetic, weak nuclear, strong nuclear) under one set of laws—physicists suspect that during the infancy of our universe these forces were inseparable and indistinguishable. Since many black holes are frozen fossils of those early moments, they must still sustain the conditions under which these forces were unified.

The Birth and Death of the Universe

All mythologies and most religions tell a story about the birth of the universe. Colorful, fascinating, and often violent, they all include a sequence of events from chaos to order, initiated and supervised by one or many gods.

Our century has made its own contribution to the list of cosmological theories. Most of these accord with the big-bang or steady-state theory. The latter proposes that our universe is in a constant and homogeneous state, governed by the same laws everywhere. Changes such as the birth and death of stars and galaxies can take place, but the overall features of the universe remain the same at all times. But since the universe is expanding, as verified by Hubble's law (see page 180), the steady-state theory faces an unresolved problem. If the universe is to remain homogeneous, the gaps caused by the expansion must be filled with fresh matter. If this is so, then, aside from the obvious question of where this matter comes from, new galaxies should be formed next to old ones. Since this has not been observed, the steady-state theory is weakened.

The big-bang theory, in contrast, remains popular and well substantiated by observation. According to this theory, 13 billion years ago all the energy and matter of the universe were packed into a space smaller than that of an atom. The conditions were such that the four major forces were indistinguishable.

Using the laws of physics, including Einstein's theory of relativity, and extrapolating current conditions back in time, theoretical physicists have developed a model of the universe in its first moments of formation. Although mind-boggling, this crude model is the best picture that science can come up with today. The most staggering part of the history of the universe are these first moments. The diameter of the minuscule cosmic seed out of which our universe was born has been calculated to be 10^{-28} centimeter (1 divided by 10,000,000,000,000,000,000,000,000,000). Its temperature was 10^{32}°C (100,000,000,000,000,000,000,000,000,000,000). It all started with a gigantic explosion. Within 10^{-37} second, the universe grew to the size of our solar system, the four major forces gained their individual identity, and the temperature dropped to 10^{13}°C. Energy was converted into electrons, protons, and neutrons, the building blocks of all matter, which later combined to form atoms. But while matter was being produced, so was antimatter, the mirror image of matter, having opposite electrical charges. (For example, the antiparticle of an electron is a positively charged electron, known as a positron.) Since any contact between matter and antimatter results in the annihilation of both, a material "civil war" took place, converting matter and antimatter back into energy. Fortunately, it appears that initially there was more matter than antimatter, and the leftover extra matter is what makes the universe today.

At about the same time, vast amounts of neutrinos were generated. These strange particles, which are continuously made in the stars, are still a mystery to physicists. We still do not know whether they have any mass. Our difficulties in understanding them are due to their strange behavior. They travel at the speed of light, they have no electric charge, and therefore they hardly interact with matter. They penetrate everything. Billions of them go through our bodies and right through the earth each second. There are more of them in the universe than anything else. One day the destiny of our universe will depend on them, as we will see later in this chapter.

Three minutes after the initial blast, the universe had cooled down to 10^{9}°C, a temperature that allows the congealing of elementary particles into atomic nuclei, like those of hydrogen and helium. These early atomic seeds were still naked of any electrons, because at these high temperatures electrons move too fast to be trapped into an atomic orbit. But ten thousand years later, when the temperatures were down to 3,000°C, slow electrons could fall into orbits around nuclei to form primeval atoms. Up to this time, matter and radiation were inseparable in a "foggy" cosmic oven. Now radiation separated from matter, and light traveling through space gave meaning for the first time to the words "brightness" and "darkness." This early radiation of 3,000°C, unleashed into an expanding universe, cooled off slowly.

In 1964 Robert W. Wilson (1936–) and Arno A. Penzias (1933–), of the Bell Laboratories, discovered this cosmic background radiation, for which they received a Nobel Prize in 1978. Trying to improve global communications, they pointed a horn-shaped antenna at the sky. They were puzzled by a mysterious background radiation that was detectable no matter when the reading was taken or how the antenna was aimed. At the same time, quite independently, a team of theoretical physicists at Princeton calculated that radiation of about 3 K (kelvins) should exist everywhere in the universe. This radiation is the cooled fading of the big bang. Like a giant energy echo, it bears testimony to the titanic explosion that took place 13 billion years ago. This discovery was the second and most powerful observational evidence in support of the big-bang theory.

Edwin Hubble (1889–1953) had offered the first evidence in 1920. His law is based on the red shift of the Doppler effect: as a source of white light recedes from the observer, it appears redder. Spectral analysis of stars and galaxies showed systematic shifting toward the red. Hubble concluded that all stars and galaxies are moving away from each other, just like marks on a balloon that is inflated. Another interesting observation that Hubble made is that the farther the star, the larger the red shift, and therefore the faster its speed. Hubble calculated the speed and distance of faraway stars and galaxies and found a linear relationship. For example, if the distance of a star is doubled, so is its speed.

The discovery of the cosmic background radiation of 3 K and Hubble's law are strong observational support for the big-bang theory. But the theory is still very crude and filled with many unanswered questions. To the question of why the universe did not collapse back into itself under its own gravity, physicists have come up with the "inflationary epoch," a split-second period during which a condition of negative gravity prevailed to inflate the universe. We still have no answers to what caused the big bang or what it was like before zero time.

Fascinating as these questions may be, we find equal excitement by asking questions about the death of our universe. Will it continue to expand forever? If not, what will happen when the expansion stops? The answers to these questions depend on how much matter exists in our universe.

When we throw a stone up, it slows down, stops momentarily, and then falls back to earth. If we increase its launching speed, it can rise higher and higher, where gravity is weaker. At some critical speed the stone will not return to earth but will continue its journey into space. This speed is known as escape velocity. For earth, it is eleven kilometers per second. Each planet has a different escape velocity because it has different mass, which in turn makes its gravitational field different. The critical parameters therefore to the question as to whether the universe will expand forever or stop are the total mass of the universe and the speed of its expansion. If there is not enough mass, the universe will continue to expand forever, diluting its energy into larger and larger space and eventually dying as a frozen dark cosmos. Such a universe is known as "open."

If, on the other hand, the mass of the universe is large enough to slow down and finally hold the expansion, then it will one day fall back into its origin in a big crunch. This kind of universe is known as "closed," and the model is often referred to as the big-bang big-crunch model. If, of course, the process repeats itself, we have an infinite number of bangs and crunches in a universe known as "oscillating."

What happens therefore depends on the total mass of the universe. We can make a rough estimate by adding the mass of all visible matter in the universe. It was found that this can provide only 10 percent of the gravity needed to hold back the expansion. Even the addition of invisible objects, such as black holes and X-ray and infrared stars, could not add up to the mass needed for a closed universe. There is one more possibility, which depends on whether neutrinos have mass. There are as many big-bang neutrinos in the universe as there are photons, which are the massless particles of radiation. Even if it takes 100 million of these neutrinos to equal the mass of one atom, they could still make 90 percent of the mass of a universe with enough gravity to slow down the expansion. Finally, such a universe will collapse back into its original size. But to measure the mass of a neutrino we must first capture one, or at least examine its interaction with matter.

The problem is that neutrinos hardly interact with matter. They go right through it as if they were particles from another dimension. Exotic experiments to detect big-bang neutrinos include the suspension of spherical sensors in the Pacific Ocean and a 400,000-liter underground tank filled with cleaning fluid (any collision with a neutrino will transform a cleaning-fluid atom into another element). Computers fed with hundreds of physics equations have developed a variety of models for the universe. Some of these show a universe that is built on a massive skeleton of neutrino matter. Others suggest that the dark empty spaces inside galaxies may be large chunks of neutrino matter.

A more "organic" and natural death of the universe has been proposed lately. It is somewhat similar to the radioactive decay process. Early studies of radioactivity showed how one material element can decay into another with a characteristic life-span, known in physics as lifetime. All elements, regardless of their apparent stability, eventually decay into other elements and into energy. The only particle of matter thought to be stable forever was the proton. Now that too is under question. Experiments are under way to detect protons decaying directly into energy. The time needed for such a decay was calculated to be so long as to be beyond our human comprehension. It is, however, estimated that if there were 10^{28} protons, approximately the number of protons in the human body, it would take two hundred years for only one of them to decay. Despite this slow process, there would be a time when all matter would ultimately be "vaporized" into energy. No theory has come forward to speculate what would happen after that.

Life Beyond Earth

Of all the marvels of the universe, the miracle of life stands out as the most highly organized and complex process. A sense of admiration and wonder develops when we realize that a tiny amount of matter on this earth developed ways by which it can reproduce, move about, even think and reflect upon itself and its environment. Could this have happened anywhere else in the universe?

"To consider the earth as the only populated world in infinite space is as absurd as to assert that on a vast plain only one stalk of grain will grow."[24] These are the words of Metrodoros of Chios, a fourth-century-B.C. Greek philosopher.

Are we alone in this vast cosmos, or are we merely a tiny part of a large intelligent cosmic family? Any attempts to answer these questions require a knowledge of the conditions necessary for the development of life on earth. Only after a thorough understanding of these conditions, as well as the processes that lead to life, can we speculate about life elsewhere.

Life on earth can be traced to its origin through the study of fossils. The accuracy of determining the age of rocks and fossils has improved through the technique of radioactive dating. This is based on the phenomenon of radioactive decay, in which an unstable element, the "parent," transforms itself into another element, the "daughter." The age of a specimen can be calculated from the measured ratio of parent to daughter elements in the sample. The age of the earth was calculated to be about 4.6 billion years, and the earliest time that plant-type life was detected is about 570 million years ago. Such long time scales are essential for the formation and evolution of life.

Past theories assumed a much shorter history for earth. For example, through the study of religious texts, Archbishop Ussher (1581–1656) of Ireland concluded that the earth was created precisely at 9:00 A.M. on October 23, 4004 B.C. An earth that is so young cannot possibly accommodate major evolutionary changes. The popular belief, therefore, was that life today is exactly as it was at the moment of creation. Geologists of the eighteenth century found, however, that the rock evidence for a much older earth was too strong to discard. The geological evidence of a much older earth, as presented by Charles Lyell (1797–1875) in his 1833 book *Principles of Geology*, provided Darwin with the time scale required for his theory of evolution.

Three types of evolution led to the development of humans. First there was a physical evolution, during which the earth, hot and with no atmosphere, slowly cooled off to form a thin crust covered by a poisonous atmosphere of methane, carbon dioxide, and nitrogen. Volcanic eruptions contributed steam and dust particles essential for rains that eventually formed the oceans. The second evolution was chemical, during which violent electric storms and penetrating ultraviolet radiation slowly converted the primitive atmosphere into what it is today. The third type—biological evolution—has obscure beginnings, but two theories have been proposed.

One, demonstrated by the classic experiment of Stanley Miller (1930–) in 1953, suggests that when the composite gases of the primitive atmosphere are subjected to lightning and ultraviolet radiation they can produce amino acids, the building block of all life forms.

Another theory, first introduced by John Rayleigh (1842–1919) in the nineteenth century, proposes that life was transplanted here by "chemical seeds" that were carried inside meteorites. The first indication that this might be true came in the early sixties from scientists who claimed to have discovered organic molecules in meteorites. These results were explained by skeptics as due to the contamination of meteorites from the earthly environment. But a team of scientists from NASA who analyzed a meteorite soon after its fall in 1969 found that it contained five amino acids.

The only conclusion that can be drawn from the meteorite evidence is that the universe may be a chemical factory that manufactures continuously the chemical seeds of life.

Mars is a planet long ago thought to accommodate life. In 1976 NASA launched the *Viking* mission to Mars. This was the first attempt to find life beyond earth. The two spacecraft that landed on Mars analyzed its soil and looked for microbes and chemicals of life. The results were disappointing. Nothing exciting was found. The discovery of active volcanoes and dry canals made scientists speculate that the planet may have undergone extreme oscillations of climate. Others suggest that the instruments were not sensitive enough, that the criteria of Martian life may be different from those of earthly life, or that any life on Mars must be living below the surface for more protection.

The *Viking* missions may have been a disappointment to some, but to others they were an important and necessary step. The human race, driven by a mysterious inner impulse, continues to search for life beyond this planet. But the formidable distances between earth and the stars make journeys to those places unlikely and impractical. Physical restrictions, therefore, force our methods to be limited in range and rather passive in nature. Realistically, our best hope for checking for any extraterrestrial intelligent life is to tune our radio ears to the galactic radio chatter, anticipating the day when we hear a faint but unmistakable intelligent signal.

What are the prospects of finding in our galaxy intelligent life that has the ability and means for interstellar communications? Radio astronomer Frank Drake (1930–), who made the first attempt to detect artificial signals from space in 1960, developed a formula that predicts the number of interstellar communicative societies in our galaxy in terms of probability factors. These factors are grouped in astronomical, biological, and sociological categories. The astronomical category contains three such factors. The first is the average annual birthrate of stars in our galaxy, estimated to be about two hundred. The second is the probability that some of these stars have planets, and the third is that some of these planets have conditions suitable for life such as ours on earth. These conditions include temperature range, atmospheric composition, and presence of water—all of which depend on the size of the planet and its distance from its star. The zone around the star in which such life conditions may be found is known as the ecosphere. These zones differ from star to star and depend primarily on the mass and temperature of the star.

The second category of probabilities is biological: first, that life has probably evolved on those planets with suitable conditions; second, that one of all those life forms has advanced enough in intelligence to have the physical and mental capabilities for interstellar communications.

Finally, the third category is made up of two sociobiological probability factors. One deals with the prospect that such an advanced intelligent form has the social need, curiosity, and desire to build such an interstellar communications system. The last probability factor hangs like a Damoclean sword over the human race. How long will we survive? This probability factor includes self-annihilation and any kind of massive natural disaster.

When all these probabilities are multiplied, we get the number of galactic intelligent societies with interstellar communications capabilities. For the Milky Way, our own galaxy of some 200 billion stars, the Drake formula predicts at least one million such civilizations. The problem is that they will be so far apart that the average time for a signal transmission is estimated to be six hundred years.

Unintentionally, we have been sending signals into space for over sixty years now. These began when the first regular broadcasting stations were put into operation in 1921. Radio signals are constantly leaking through the ionosphere, the thin electrified layer of the atmosphere that acts as a giant radio reflector. Since we are still infants in a technological sense, our signals are poor and haven't had the time to go far into space. It is therefore best if we concentrate our efforts on listening for signals that may have been sent many years ago by faraway advanced civilizations. To many this may sound like a fruitless exercise, yet the day will come when an unmistakable intelligent whisper from outer space will generate shock waves of excitement in the hearts of all human beings. This event will be marked as the greatest single discovery in human history. Its powerful effect on the human condition is difficult to forecast. We only have to remind ourselves of the excitement caused by the launching of *Sputnik* in 1957 and the discovery of pulsars in 1968.

To ensure success in detecting extraterrestrial intelligent signals, certain technical problems must first be overcome. A major question is to which frequency we should tune our radio antennas. A radio signal that travels with the speed of light may be sent in any one of an infinite number of frequencies. Laboratory tests have shown that only some frequencies have the ability to penetrate the hydrogen clouds and interstellar dust found in space. Intelligent extraterrestrials should be aware of this, too, choosing therefore

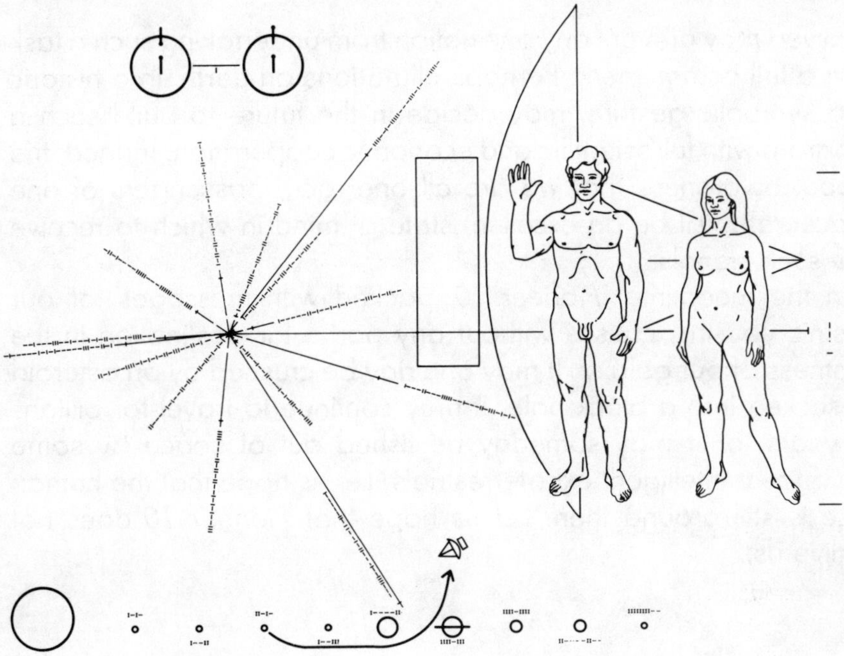

The plaque placed aboard *Pioneer 10*, the first spacecraft to leave our solar system. (Courtesy Dr. Carl Sagan)

one of the penetrating frequencies for their cosmic radio transmissions. But even these are too many, so the dilemma remains.

New technological innovations may soon provide the solution to this problem. Through a sophisticated combination of computers and electronics, it may be possible to tune to a million different frequencies simultaneously. It will be like listening to a million radio stations at once. But other problems still remain, such as to which part of the sky the radio antennas should be pointing and what minimum strength signal our receivers will be able to detect.

Project Cyclops is a multibillion-dollar proposed project that may offer the best chance of detecting an extraterrestrial intelligent signal. Consisting of fifteen hundred 100-meter dishes, it will spread over an area of two hundred square kilometers, scanning most of the sky at many frequencies. It will be able to detect radio signals sent from as far away as one thousand light-years or others that may possibly have leaked from planets with intelligent civilizations up to one hundred light-years away. But the prohibitive costs involved may prevent any one nation from undertaking such a task with a full commitment. Perhaps all nations on earth, in a historic and symbolic gesture, may decide in the future to build such a program with full scientific and economic cooperation. Indeed, the global awareness that we are all one race, passengers of one spacecraft, will be an excellent state of mind in which to receive any such signals.

In the meantime, *Pioneer 10,* packed with messages for our cosmic cousins, cruises without any particular destination in the vastness of our galaxy. It may one day be crushed by an asteroid or sucked into a black hole. It may continue to travel for billions of years, or it may someday be fished out of space by some advanced, intelligent extraterrestrials. Let us hope that the human race is still around then. Let us hope that *Pioneer 10* does not outlive us.

Chapter Eight

Space Colonization

Exploration is as old as the human race. This inexplicable urge to explore unknown territories, derived from that blessed gift of curiosity, is much more than an adventure and a way of survival. It is an expression and a testimony to the essence of the human spirit. From the barren deserts to the swampy tropics, from mountaintops to the polar caps, each corner of our planet bears our footprints. Brave discoverers such as Marco Polo, Columbus, and James Cook risked their lives seeking new worlds, trade routes, and new cultures. They are not just the national pride of this or that country but more like the heroes of all the human race, just as are the seven American astronauts in the *Challenger* space shuttle tragedy.

The fruits of exploration are rich and many. Gold, spices, silk, plus potatoes, corn, and other fruits and vegetables are few of the material rewards of the past. The exploration of space offers its own exciting list of rewards. The extraction of precious raw materials from the moon and asteroids, abundant clean and cheap solar energy, and opportunities to manufacture new industrial and pharmaceutical products are only some. As in past explorations, the new frontiers offer the chance of establishing large and permanent colonies—in space and on the moon and the planets.

Besides the material rewards, explorations bring many other benefits of immeasurable importance. Each mission adds to our understanding of the world around us, furnishing information that can improve our chances of survival as a species. Space exploration will continue to provide us with answers to our many questions about the origins of our universe, the forces and objects that fill its space, and perhaps its final destiny.

The launching of *Sputnik 1* on October 4, 1957, signaled the beginning of a new age of exploration with an infinite potential. In only a few years after *Sputnik,* we have landed on the moon, sent spacecraft to study the planets of our solar system, and sent *Pioneer 10,* the first artifact, beyond our solar system, traveling right now toward some distant star. In only a single generation we have learned more about our universe than we did in all the centuries past. Yet we realize that this is merely the crude beginning of a new and exciting era. More is still to come.

Explorations can be highly profitable enterprises, but they require some very special kind of investments. Besides the monetary support, they require above all commitment, dedication, and courage. Explorations are not always guaranteed success, so those undertaking them must learn to cope with failure, disappointment, and criticism. True explorers rise above such setbacks, exhibiting the qualities of their generation and pushing on with their tasks. The new frontier of space is indeed a golden opportunity to test those human qualities. The new challenges of space may very well be the reason for developing into an even better species.

The Quest to the Moon

Fascination with flying goes back thousands of years. It always symbolized power, perhaps because it was seen as defying gravity and venturing into a territory that belonged only to birds and gods. Almost all mythologies include creatures with wings, gods who lived in the sky, and stories of men who flew. Greek tales, for instance, tell of Icarus and Daedalus, father and son, who flew to escape their imprisonment by wearing wings of feathers put together with wax. Though Daedalus warned his son not to fly too near the sun, the youth disobeyed. His wax wings melted, and he fell to his death.

Between 1486 and 1514 Leonardo da Vinci (1452–1519), the Italian Renaissance genius, studied extensively the flight of birds, identifying the physics laws of flying, a science known today as aerodynamics. Whereas others saw only beauty and grace in flying birds, da Vinci saw them also as ingenious flying machines. He wrote, "A bird is an instrument working according to mathematical laws. . . ."[25] He believed that the secret to man's flying was in the "science of the wings," which could be developed from the study of birds. Besides extensive observation of birds in flight, da Vinci designed and built study models of gliders, parachutes, and a rather unusual screw-shaped helicopter.

The dream of flight came true on November 21, 1783, but it was by riding a hot-air balloon rather than by imitating a bird. Designed out of paper by the French Montgolfier brothers, Joseph Michel (1740–1810) and Jacques Étienne (1745–99), the balloon was 49 feet (15 meters) in height and 1,742 pounds (790 kilograms) in weight. With its two passengers it flew over Paris for twenty-five minutes, traveling five and a half miles. Soon after that a variety of new designs made ballooning a very popular but still dangerous activity. Records were broken repeatedly: crossing natural boundaries like channels or mountaintops, staying up the longest, traveling the fastest, or covering the longest distance.

In the summer of 1878, when the Wright brothers were eleven and seven, their father gave them a flying machine, a toy helicopter powered by rubber bands. Their interest in flying grew, and as successful bicycle manufacturers later on, Wilbur (1867–1912) and Orville (1871–1948) read everything they could find on flying. They learned of Otto Lilienthal (1848–96) in Germany, who had already made hundreds of flights in his gliders. Without even high-school diplomas, the two brothers set themselves on a course to become experts in the science of flying.

Using their bicycle workshop, they tested more than two hundred types of wing surfaces and a number of gliders. With a total expenditure of less than a thousand dollars, their flying machine, a glider with a small twelve-horsepower motor and two screw propellers, was ready for testing. Their first successful flight took place near Kitty Hawk in North Carolina on December 17, 1903. It lasted only twelve seconds and covered only 120 feet, yet it went down in history as the first sustained, controlled, powered flight of a machine heavier than air.

The story of rockets starts in the thirteenth century, when the Chinese used "fire arrows" to scare away invading Mongols. Fireworks had been used much earlier by the Chinese in their festivals. When later on they were introduced in Europe, they were used crudely during celebrations and occasionally very ineffectively in battle. By 1700 these ancestral firecrackers had developed into sophisticated noisemakers, with multistage firing, stabilizing vanes, and effective designs. Despite their widespread use and sophistication, rockets did not capture any serious scientific interest.

Almost twenty years after the Wrights' historic flight, while progress in aeronautics was advancing very fast, a lone figure developed a keen scientific interest in rockets. In 1919 Dr. Robert H. Goddard (1882–1945), a young physicist who loved H. G. Wells's science-fiction novels, published a paper titled "A Method of Reaching Extreme Altitudes." Seven years later he launched the first liquid-propellant rocket. He continued his research with small grants and lots of sarcastic remarks about the futility of his work from colleagues, peers, and newspaper editorials. But Charles Lindbergh, America's aviation hero, saw the great significance of Goddard's work and rallied financial support that assured its continuation.

When, in 1950, Wernher von Braun (1912–1977), the German rocket expert who built the V-2 rocket for the Nazis, saw the 214 patents that Goddard left, his comment was "Goddard was ahead of us all." Von Braun went on to become the major architect of the American rocket program. What remains as an interesting paradox is that during World War II the Germans put a lot of emphasis on the military use of rockets, whereas the American government apparently lacked any interest in Goddard's work.

Although science and technology could not make space flight possible before 1958, the idea was deeply rooted in our culture from very early times. It was found first in mythology and later in science and literature. The first known science-fiction story was written in the second century by a Syrian lawyer-lecturer called Lucian (125?–180?). It describes a ship that was caught in a storm and lifted to the moon, where its crew witnessed a war between the inhabitants of the moon and the sun.

Johannes Kepler (1571–1630), the famous seventeenth-century astronomer, described accurately the physical and physiological changes that an astronaut traveling from the earth to the moon would experience. Even the great Newton launched in his mind the very first artificial satellite, as described in his famous book *Principia.* The Russian space pioneer Konstantin Tsiolkovsky (1857–1935) described accurately the panoramic view of the earth from an imaginary orbiting satellite. But undoubtedly the most sensational and accurate science-fiction writer was Jules Vernes (1828–1905), whose story of a journey to the moon and back was not only fascinating but also an accurate description of things to come. Launched from an enormous cannon, the capsule had to attain the escape velocity, the twenty-five thousand miles per hour needed for any object to escape the earth's gravity. The passengers floated in near zero gravity, and the spacecraft became white-hot from friction during its reentry into the atmosphere.

Fantasy became reality on October 4, 1957, when the Soviets launched the first satellite, which signaled the birth of the space age. Millions of people throughout the world listened with fascination, pride, and even fear to the electronic blips sent out by this football-size 184-pound electronic sphere orbiting the earth every ninety-six minutes. It is unfortunate that about a year earlier an American-launched rocket was purposely kept short of reaching orbital velocity. The reason for this was disclosed later on to be fear in the Department of Defense that the army might launch an unauthorized satellite. Despite later excuses, apologies, and cover-ups, one thing remained sure in the eyes of the American public: that America missed being first in a new epoch and that it was behind in the space race.

Four months after *Sputnik,* the United States launched *Explorer,* the first American satellite to make a major discovery of the space age: the Van Allen radiation belts surrounding earth. In the meantime, the feeling of urgent need of an agency that could restore pride in American space science was growing stronger. On July 29, 1958, President Eisenhower signed into law the birth of the National Aeronautic and Space Administration (NASA), "devoted to peaceful purposes for the benefit of all mankind."

Early skepticism about the practical use of the space program soon disappeared. The early use of satellites revolutionized communications and weather forecasting. Larger and more sophisticated satellites later on would photograph every inch of our planet in visible, ultraviolet, or infrared light. These satellite photographs were filled with valuable information about the earth's resources. Satellites were later used in a multitude of projects in urban development, agriculture, pollution control, forestry, and meteorology, as well as gathering valuable military information.

When President John F. Kennedy delivered his special message to Congress on May 25, 1961, he made on behalf of the American people a bold commitment to the space program. His words remain in history.

> I believe that this Nation should commit itself to achieving the goal, before this decade is out, of landing a man on the moon and returning him safely to earth. No single space project in this period will be more exciting, or more impressive to mankind, or more important for the long-range exploration of space; and none will be so difficult or expensive to accomplish.... But in a very real sense, it will not be one man going to the moon—we make this judgment affirmatively—it will be an entire nation. For all of us must work to put him there.

The Congress approved and the American public enthusiastically supported the expanded and ambitious long-term space exploration program. About that time, too, NASA was completing Mercury, its first major project, with solo pilots. Its objectives were to orbit a manned spacecraft around the earth, to investigate human ability to function in space, and to recover both pilot and spacecraft safely. All objectives were met, with John Glenn becoming the first American sent into orbit, on February 20, 1962. The Russian cosmonaut Yuri Gagarin was the first to orbit the earth, on April 12, 1961. In this historic flight Gagarin used *Vostok,* a five-ton spacecraft that made a single orbit at 188 miles in one hour and forty-eight minutes.

The second U.S. manned space program was Project Gemini, which involved twelve flights, including two unmanned flight tests of equipment. Carrying two pilots in each mission, this project provided experience of how to live and maneuver in space. The major objectives were clear-cut: to subject astronauts and equipment to space flight up to two weeks in duration; to rendezvous and dock with other orbiting vehicles and maneuver the docked combination; to perfect methods of entering the atmosphere and landing at a preselected point on land. All goals were met except for the last, which was canceled in 1964.

"That's one small step for a man, one giant leap for mankind." Uttering these historic words while being watched on millions of television sets around the world, Neil Armstrong stepped onto the

Neil Armstrong, the first human on the moon. (Photographs in this chapter courtesy NASA)

lunar surface on July 21, 1969. "We came in peace for all mankind," read the inscription on the unveiled plaque. This was the climax of perhaps the most exciting NASA program—the Apollo project.

But Apollo's goals went beyond landing Americans on the moon and returning them safely to earth. Based on the cumulative experience of Mercury and Gemini and using the mighty Saturn rocket as its launch vehicle, the Apollo program established preeminence in space for the United States, carried out a scientific exploration of the moon, and developed the technology needed for humans to work on the lunar environment. Considered one of the most extraordinary technological and scientific accomplishments, the Apollo project was also a bridge to the *Apollo-Soyuz* project and *Skylab*.

Skylab was America's first experimental space station. Designed for long missions, *Skylab* had two basic objectives: to prove that humans could live and work in space for extended periods and to expand our knowledge of solar astronomy. Three three-person crews occupied it for a total period of 171 days and carried out nearly three hundred experiments, including medical experiments on humans. *Skylab* remained in orbit for more than six years, completing 34,981 orbits. It returned to earth on July 11, 1979, and disintegrated in the atmosphere, scattering debris over the Indian Ocean and the deserts of western Australia.

The Space Shuttle

Even before the launching of *Sputnik* scientists and engineers recognized the need for a versatile and reusable vehicle if space was to be explored regularly and economically. Unfortunately, the technology required to build such a spacecraft was not available then. Besides, the competitive race with the Soviets for larger and more accurate ballistic missiles kept the emphasis on improving conventional rocketry rather than experimenting with new designs.

The spectacular success of the Apollo moon landings boosted scientists' confidence about the power of space science and technology. This in turn rekindled interest in the next challenge in the space program, that of cost reduction. On January 5, 1972, President Nixon stated:

> The United States should proceed at once with the development of an entirely new type of space transportation system designed to help transform the space frontier of the 1970s into familiar territory, easily accessible for human endeavor in the 1980s and '90s.... It will revolutionize transportation into near space by routinizing it.... It will take the astronomical costs out of astronautics.... This is why commitment to the Space Shuttle's program is the right next step for America to take.

On March 15, 1972, after years of extensive studies that included aspects of science, engineering, safety, and economics, NASA decided on a three-part configuration, the Space Transportation System (STS). This is better known as the space shuttle, which consists of an expendable external tank, two partly reusable solid rocket boosters, and a reusable orbiter. The shuttle is put together inside the Vehicle Assembly Building, a 480-feet-tall superstructure that was originally built and used for assembling the Saturn-Apollo vehicles for the lunar program. Here the orbiter is prepared and then lifted to a vertical position and bolted to the external tank, which is already joined with the booster rockets. The mobile launching platform, upon which the shuttle is assembled, is then carried at one mile per hour by four enormous crawler tracks to the launch site, a little over four miles away.

The solid rocket boosters are 150 feet high and 12 feet in diameter, and they contain a total of a million kilograms of solid fuel. This is a mixture of aluminum powders, iron, and a polymer. The fuel looks and feels very much like a hard rubber typewriter eraser. It is very safe because it cannot be ignited by static electricity, friction, or impact. When ignited by the rocket motor, it burns at 3,200°C, generating a thrust of 5.5 million pounds. The nozzles of these rockets can be swiveled hydraulically by the orbiter's computer to control the direction of the thrust.

After firing, the rocket boosters burn for two minutes and seven seconds, then are separated from the external tank by small ex-

The space shuttle carried by the mobile launcher platform from the assembly building to the launch area.

plosive devices. They fall toward the earth for four minutes, reaching a speed of 3,000 mph, which is then reduced to 60 mph through a succession of parachutes. Floating vertically in the Atlantic Ocean, they emit radio signals that are picked up by recovery vessels. Air is pumped into them so that they can float horizontally while being towed ashore. After a thorough cleaning, inspection, and refurbishing, the rockets are repacked with fuel and are ready for their next flight.

The external tank stands as tall as a fifteen-story building (154 feet) and as wide as a farm silo (27.5 feet). It is filled with fuel—liquid oxygen and liquid hydrogen—two and a half hours before lift-off. The fuel is for the shuttle's three main engines, which are in the stern of the orbiter. Each of these engines is 14 feet tall and 7.9 feet in diameter, producing 375,000 pounds of thrust—equiv-

Challenger bolted to its launch rockets. For scale, notice the technicians near its engines.

Columbia, bolted atop a modified 747 jumbo jet, returns home to Florida after landing in California.

alent to about 6.5 million horsepower. Computers in the orbiter control the rate of fuel consumption and, by swiveling the engines, the angle of thrust. After about eight minutes from lift-off, the empty external tank is jettisoned; unlike the rocket boosters, it disintegrates as it reenters the atmosphere, with any remaining fragments falling into a remote ocean area.

After its separation from the external tank, the orbiter coasts freely for a short time. Then it fires two smaller engines, which are fueled by its own internal tanks. In two more minutes, the orbiter reaches its required orbital velocity of 17,500 mph. With a few more firing adjustments, it finally establishes its permanent orbit. Almost an hour after lift-off and while going over Australia, a pair of clamshell doors open outward to expose the cargo bay. These doors incorporate radiators that dissipate heat generated by the electrical equipment.

The space shuttle orbiter is the most sophisticated flying machine ever built, standing as a symbol of power for twentieth-century

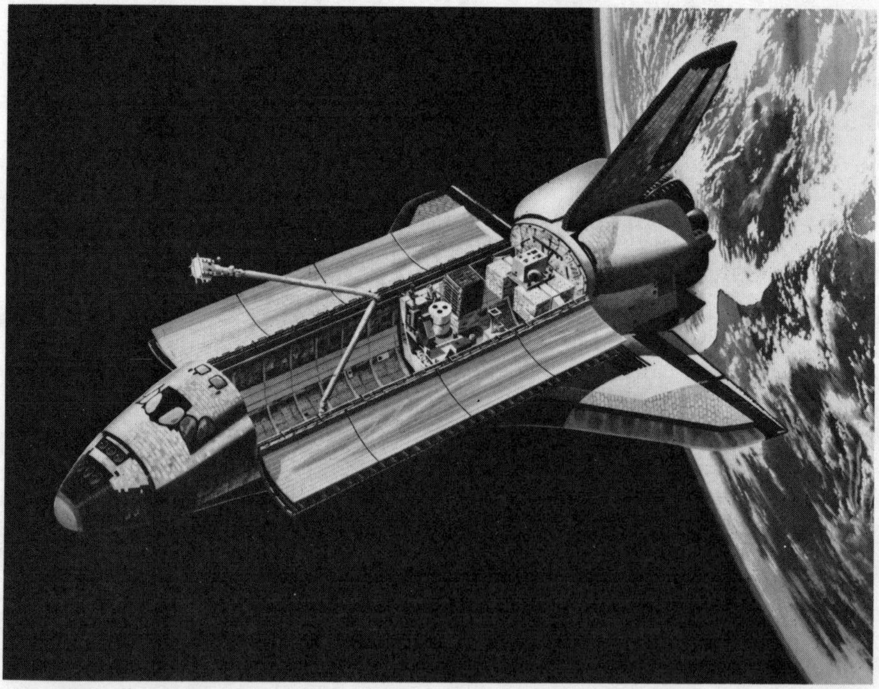

An artist's illustration shows *Columbia* in orbit, deploying one of its experiments.

science and technology. It takes off like a rocket, orbits the earth like a spacecraft, and lands like a glider. Its delta-wing-shaped body is 121 feet long and 79 feet wide, with a weight of 166,000 pounds. It has been subjected to extensive engineering design and testing, and in its final version it is packed with the latest in technology. It has forty-nine engines, twenty-three antennas, five computers, and a complete support system that draws its electricity from generators that also produce drinking water. Designed to be reused for up to one hundred missions, the orbiter can stay in orbit for seven to thirty days at a time, carrying a maximum crew of seven.

The orbiter's thick body is covered with a very special type of insulation that protects it from the searing heat generated by friction with the atmosphere during reentry. Earlier spacecraft used thermal shielding that would peel off gradually as it got hot at reentry. The insulation for the orbiter, however, stays intact and must last for many missions before replacement. It is made of a light material that looks and feels like styrofoam but has some amazing insulating properties. It can be red-hot on one side while remaining cold on the other. It took thirty-two thousand tiles, each individually contoured and custom-built, to completely cover *Columbia,* the first orbiter. It is not a surprise that it was often referred to as "the flying brickyard." The number of tiles was reduced later by using insulation "blankets" to cover large sections of the orbiter.

The orbiter draws its internal power from separate electrical and hydraulic systems. Hydraulic power, used for wing flaps, rudder–speed brake, main engine valves, landing gear, and wheel brakes, is generated by three gas turbine pumps that are driven by the decomposition of hydrazine (a clear oily liquid fuel that smells like ammonia). Electrical power is generated by the electrochemical reaction of hydrogen and oxygen in three fuel cells. A valuable by-product of this is drinking water. The electrical consumption varies from an average of 14 kilowatts to a peak of 36 kilowatts when most equipment is in operation during orbit.

The crew and passengers occupy a two-level cabin at the forward section of the orbiter. They experience a maximum gravity load of only 3 g (g is the unit for normal gravity on earth) during lift-off, and less than 1.5 g during reentry. These forces are due

Columbia's rear wheels touch the ground at Edwards Air Force Base to successfully complete a ten-day mission.

STS-4 thunders away from Launch Pad 39A for a seven-day earth orbital mission and the final development flight for the Space Transportation System.

Columbia lifts off on April 12, 1981—the dawn of a new age in space flight.

to acceleration and are designed to be about one third the levels experienced on previous manned flights. This is one of many features designed into the space shuttle that allow even those who aren't in the perfect physical condition of astronauts to be passenger-specialists in the program.

The flight deck looks like that of an airplane but with many more switches, panels, computers, and display screens that monitor and control the complex electronic system of the orbiter. The living and working quarters offer a comfortable environment with proper temperature and humidity and filters that remove odors and carbon dioxide. The few appliances on board include an oven, hot and cold water dispensers for preparing freeze-dried foods, storage for seventy-four kinds of food and twenty beverages, places for eating utensils, a water tank, and trash bins. It has a washroom and a toilet and sleeping bunks, and extra seats can be made available for a rescue mission.

At the back of the cabin there is a cylindrical compartment, with airtight hatches on oposite ends, that leads into the cargo bay. To work in the cargo bay, astronauts have to wear their space suits with backpacks filled with oxygen and a cooling system for up to six hours of EVA—extravehicular activity. When necessary, they also wear a hand-controlled back rocket kit that enables them to fly outside the cargo bay for special and delicate operations. EVA tasks may include inspecting and photographing payloads, installing or removing film cassettes, repairing, replacing, or calibrating instruments, performing experiments, unloading cargo, retrieving satellites, and carrying out overall inspection and repair of the orbiter.

An interesting feature of the cargo bay is the manipulating arm. It takes six electric motors to operate this fifty-feet-long giant, which has shoulder, elbow, and wrist joints like the human arm. Remotely controlled from the orbiter flight deck and guided by its own TV camera and lights, the arm can lift out or haul into the cargo bay satellites or materials and supplies for the space station.

There are many other unique features of the shuttle besides its great advantages of economy, reusability, and versatility. It can be launched on relatively short notice, so that it can be used for rescue missions in space. It can be used for special urgent missions

Astronaut Bruce McCandless II reaches a maximum distance from *Challenger* before reversing direction of his manned maneuvering unit and returning to *Challenger*.

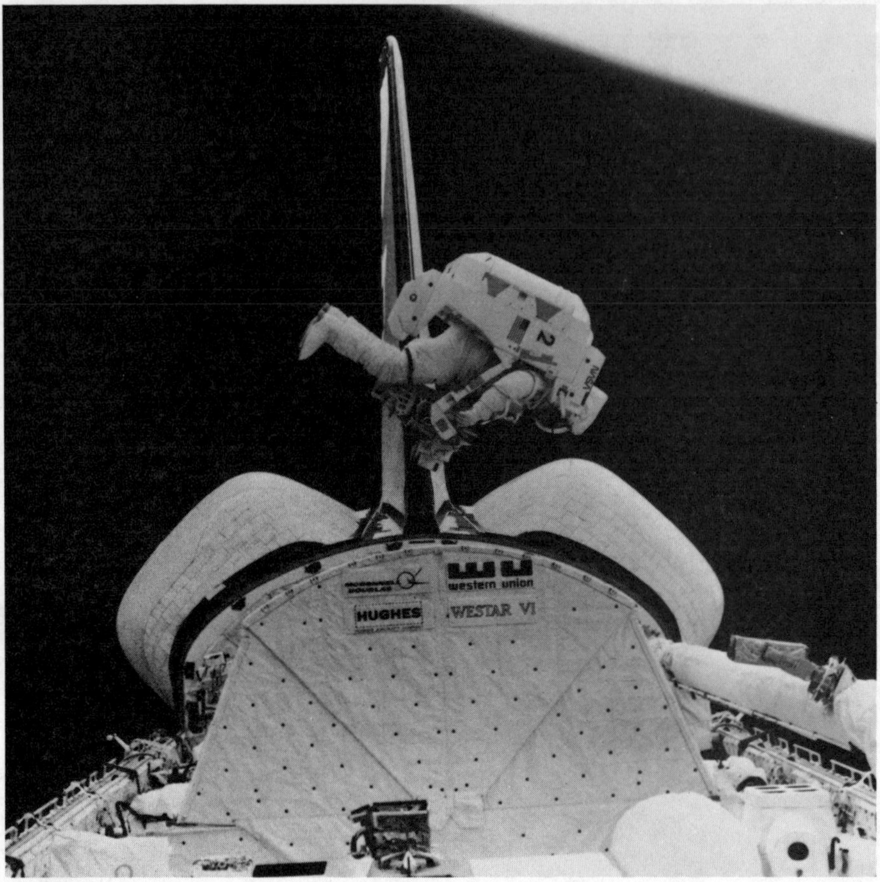

McCandless uses a jet-powered, hand-controlled maneuvering unit in a space shuttle flight test.

to monitor global emergencies like floods, fires, or crop blight. Unlike previous manned spacecraft that splashed in the ocean, requiring a costly recovery, the shuttle lands like a glider. Its low acceleration allows scientists and engineers to ride as passengers with only a few weeks of training.

Since the first space shuttle flight in April 12, 1981—STS1 *Columbia,* with John Young and Robert Crippen as crew—the shuttle program has come a long way. It carried out successfully many tests and observations, and its cargo experiments spanned the scientific disciplines. Satellites were launched and retrieved, and astronauts "walked" in space wearing their amazing backpacks.

Almost every mission had its own minor setback, such as failure in satellite launchings, malfunctions of equipment, and the loss of

An artist's concept of how a space-suited astronaut retrieves a satellite.

The remote manipulator system arm of the space shuttle.

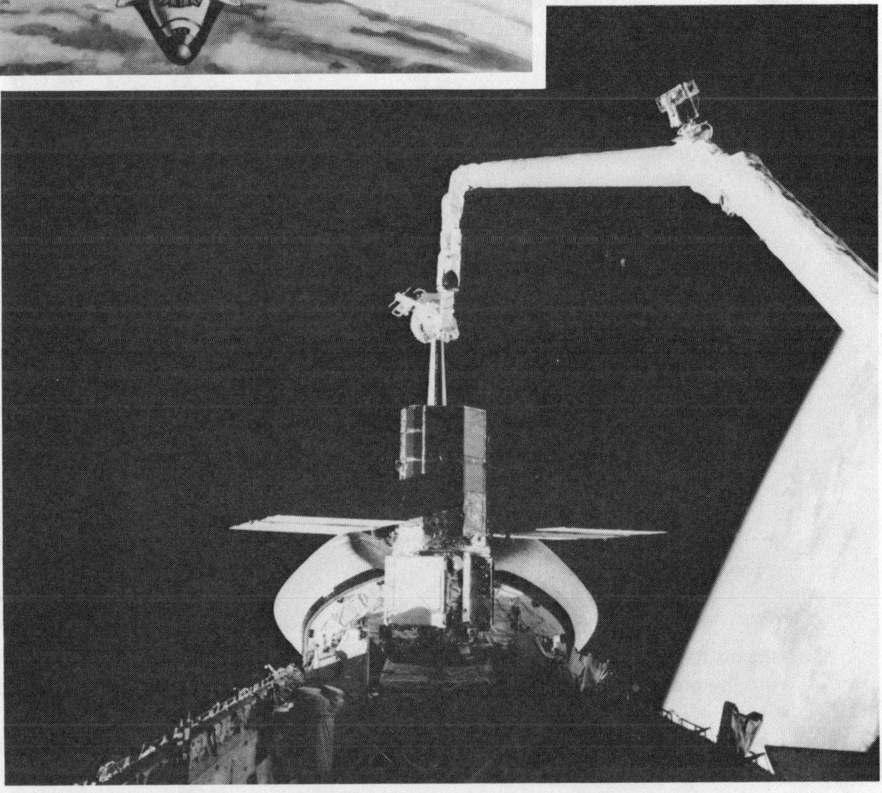

thermal tiles. The most dramatic setback was the tragic midair destruction of *Challenger,* seventy-three seconds after its lift-off on January 28, 1986. Its seven occupants, six astronauts and the first civilian passenger, a teacher, were killed. The cause of the disaster was apparently a faulty seal on the right rocket booster, damaged either during assembly or by freezing weather the night before. Economic and political pressures put on NASA to perform efficiently and on schedule may have contributed to this tragedy. But the destruction of *Challenger* is much more than the loss of over a billion dollars in sophisticated technology and the inevitable delays that follow such accidents. It serves as a permanent reminder that daring projects such as space exploration will always be accompanied by dangers, risks, and even more sacrifices no matter how careful we are. Despite such setbacks, we choose to go on, regarding the shuttle program as an indispensable key to the exciting horizons that lie ahead.

Astronauts John W. Young (left), crew commander, and Robert L. Crippen, pilot, were the first space shuttle astronauts as the crew on the maiden voyage of *Columbia*.

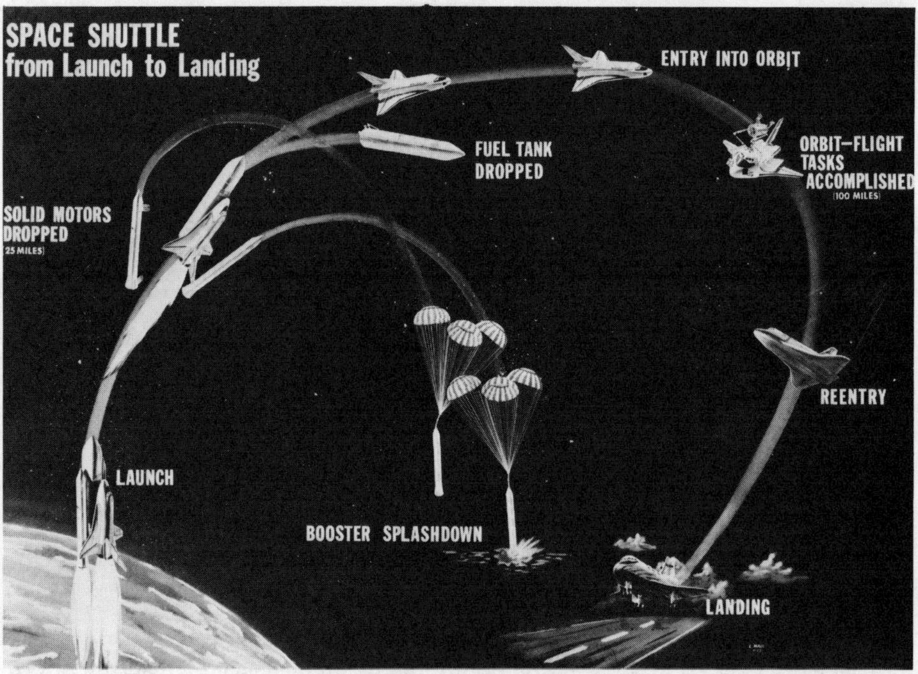

The space shuttle from launch to landing: the critical stages of a typical mission.

The many future missions, already booked years in advance, are some of the most daring tasks ever set up. Of all these projects, four stand out: Spacelab, the space telescope, the launching of SETS (Solar Energy Transportation Systems) spacecraft, and, the most exciting of all, the building of a permanent space station.

About a fourth of all space shuttle missions will include Spacelab, a versatile laboratory designed and constructed by the nations of the European Space Agency (ESA) and operated by NASA. The ten member countries are Austria, Belgium, Denmark, West Germany, France, Italy, the Netherlands, Spain, Switzerland, and the United Kingdom. This international space project reflects NASA's commitment to cooperate with other nations in peaceful activities that benefit all, as mentioned in the 1958 congressional act.

Spacelab is made of a cylindrical pressurized capsule that fits inside the cargo bay of the orbiter. Like the shuttle itself, Spacelab is reusable, designed to be launched and returned with the orbiter as many as fifty times during its ten-year life expectancy.

The official STS-9 mission emblem, the first mission of the European-built Spacelab, with more than 70 experiments on board. This was a joint venture between NASA and ESA.

Its interior is equipped with instruments, racks, and work benches, allowing three or four scientists to carry out their experiments in a comfortable shirt-sleeve environment.

A usual attachment to the capsule is an open platform, a sort of open back porch, called the pallet, which carries instruments and materials that are tested or calibrated in the environment of space. As many as five pallets can be flown at a time. There can be as many as ten different pallet-capsule combinations.

Spacelab research will often complement long-term satellite observation of the sun and solar wind, comets, and high-energy radiation from distant parts of the universe. Other studies will include the earth's electromagnetic environment and the upper atmosphere. Experiments will study the possibilities of space processing of industrial and biomedical products, and the effects of space on humans, plants, and animals.

Another major NASA project that depends heavily on the shuttle is the space telescope. Scheduled to be placed in permanent orbit

An artist's illustrations show the space shuttle carrying Spacelab-1 in its bay.

by the space shuttle in 1989, it is of paramount importance to astronomy. Having a weight of ten tons and a length of forty-six feet, this huge and powerful observatory will see more than earth instruments of the mysterious sky. Orbiting above the murky and turbulent atmosphere that blurs the view of telescopes on earth, it will use five extremely sensitive instruments to detect light from so far away that it may have taken billions of years to reach us. The study of this ancient light may solve some of the mysteries of the universe in its infancy.

Other tasks of the telescope will be to study the birth of new galaxies and to search for planets around stars like our sun. Throughout its expected twenty-year lifetime, the telescope will be maintained and repaired by astronauts from the space shuttle. Periodically it will be brought back to earth for a complete overhaul.

The study of the composition and nature of asteroids and comets will one day be necessary if we are to get information on the birth and history of our solar system. The need to fly close by these small bodies and even to collect and return samples makes it necessary to design a new type of spacecraft with a more powerful and more economical means of propulsion.

One such possible design is SETS (Solar Energy Transportation Systems), which would operate only in space and therefore could be launched only from the space shuttle. Its fuel is electricity from sunlight, collected by large solar panels. The electricity is used to convert neutral atoms, probably those of mercury, into ions (electrically charged atoms). These are then accelerated and ejected through a nozzle, providing a weak thrust to push the spacecraft forward. Although the propulsion is weak, its continuous application over weeks or months (until the supply of mercury runs out) can build up an enormous speed.

The most exciting of all future NASA projects remains, however, the building and operation of an orbiting space station. The motive behind this bold adventure is to establish colonies in space, and it is very much in resonance with some ancient human qualities: to explore, to conquer, and to settle in new lands. This new adventure will open up a vast new frontier with all its challenges, rewards, and dangers. Building colonies where the gods once lived will require much more than money. It will demand a blend of determination and courage with scientific know-how.

The Space Station

"We can follow our dreams to distant stars, living and working in space for peaceful economic and scientific gain. Tonight, I am directing NASA to develop a permanently manned Space Station and to do it within a decade." With these words in his State of the Union message on January 5, 1984, President Reagan assigned NASA its toughest project yet.

Orbiting the earth like an inhabited island, the space station will be NASA's last great venture of this century and the stepping stone toward an even more ambitious space program in the next century—the colonization of the moon and the other planets. Thus, it may be as important as all of the previous great explorations put together. It will undoubtedly become the focus of American brainpower and pride, a symbol of courage and progress, and a vivid example of the essence of the human spirit.

The space station will be built from sections prefabricated on earth and transported to orbit in the shuttle's bus-size cargo bay. There they will be unloaded and assembled by astronauts wearing space suits and propelling themselves with jet-powered backpacks. It will take about eight shuttle flights and an estimated eight billion dollars to build the prototype space station. It is destined to be up there for thirty years, orbiting the earth at an altitude of 250 miles. Early crews will be limited to eight people, but as the space station grows, it is estimated, as many as twenty workers could be accommodated. Regular shuttle flights will deliver new supplies and fresh crews every few weeks.

As currently foreseen by NASA, the crew quarters will be in a cylinder-shaped habitation module about as large as a medium-size house trailer. A similar module will be used as a laboratory. Both modules will have an atmosphere like that in an airplane's passenger cabin, and they will be connected to each other with a tunnel. Other modules of the same or different size and shape will house power-generating equipment, supplies, and reserve machines.

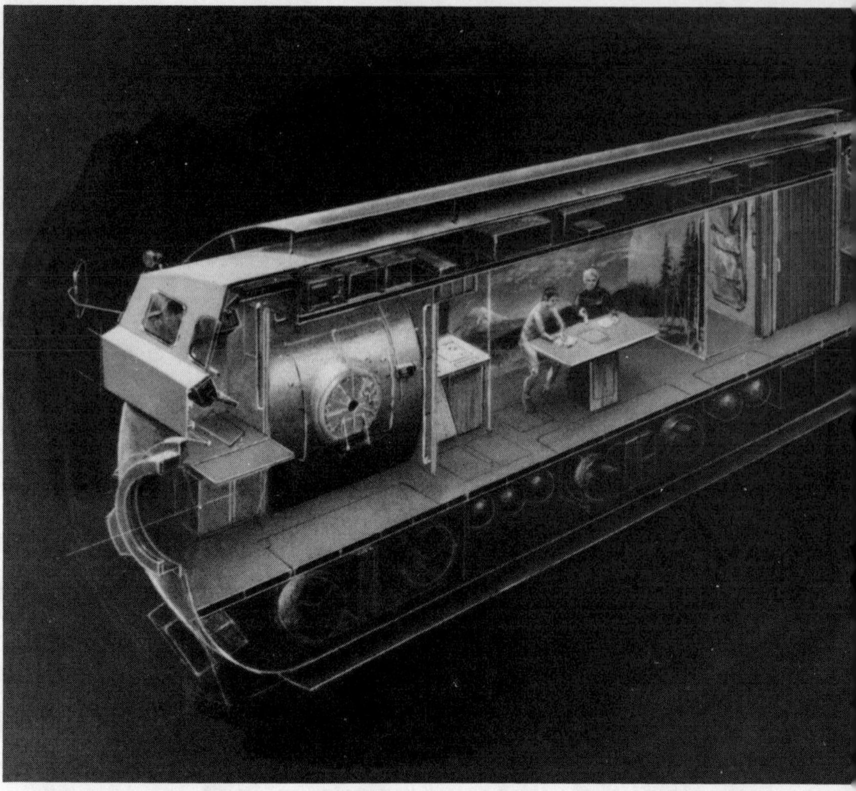

Cutaway sections of two living modules, the command and control center of a space station. The left section includes dining area, food storage facilities, sleeping quarters, and deck for off-hours observation of earth and other planets. The right section includes sleeping quarters, exercise and recreation area, laboratory, and storage area.

Walter Froehlich, the author of the NASA publication *Spacestation,* says, "Protruding like paddles from the sides of the modules will be rectangular panels covered with solar cells for power production. Other panels will carry radiators to dissipate excess heat generated from Station operations into the cold vacuum of space."

Communication antennas of various sizes and shapes will extend from the modules or other structural beams. Attached to these beams will be "seatings" for mounting scientific instruments or construction equipment. One such tool will be a giant crane, resembling a mechanical arm, that will be used for unloading the shuttle.

To be cost-effective, the space station must be as self-sufficient as possible, recycling waste into usable materials. NASA plans to

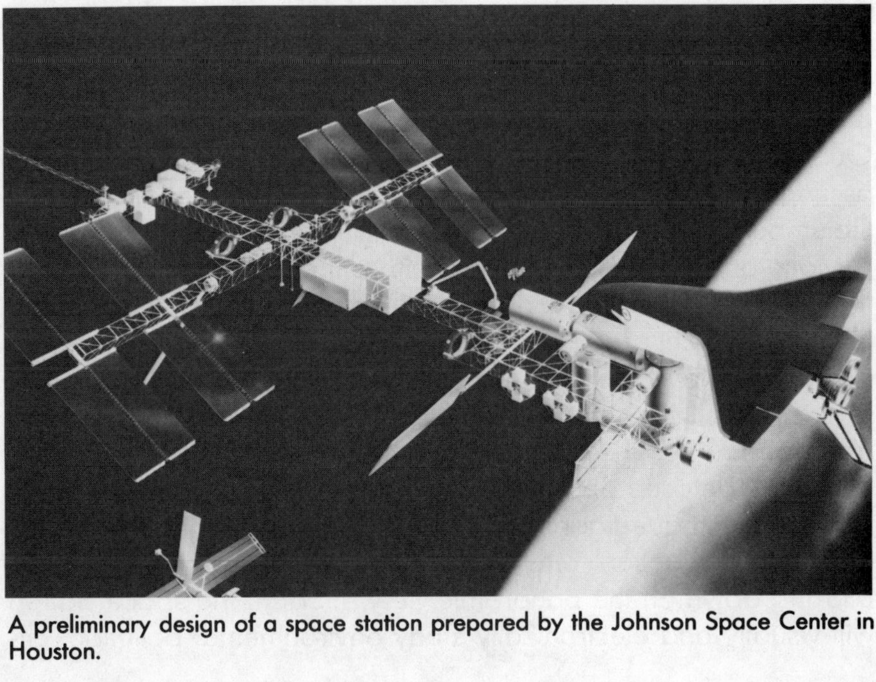

A preliminary design of a space station prepared by the Johnson Space Center in Houston.

spend millions of dollars researching this new concept, known as regenerative technology. For example, resistojets—small rockets that will fire continuously to keep the station from slowly falling to earth—may use exhaled carbon dioxide as a fuel. Plants that will be grown inside cylindrical containers (which will spin to generate artificial gravity) will also consume carbon dioxide and provide food and oxygen for the crew. New and powerful solar cells made of gallium arsenide will generate the 100 kilowatts of the station's necessary power. Machines may even manufacture trusses out of raw materials.

Opponents of the project, among them some prominent scientists, question the wisdom of spending such vast amounts of money and effort, arguing that the anticipated benefits may not be worth the investment. It is ironic that similar opposition was voiced also during the early stages of the satellite-launching era. Satellites have turned out to be financially the most profitable branch of the space program.

The greatest money-making potential of the space station comes from vacuum and zero gravity, which make it the ideal place to manufacture certain drugs, alloys, crystals, semiconductors, and even perfect ball bearings. Private industry will be able to rent research and manufacturing facilities to develop and produce new products. But even if no financial benefits were in sight, the spirit of our human nature demands that we develop the space station. Our venture out to this new frontier could very well turn out to be as unpredictably beneficial to the world as was the voyage of Columbus.

Monetary rewards alone are neither the motivating factor nor the measure of success for any genuine scientific project. Keeping loyal to this spirit of scientific investigation, the space station promises to be one of the most powerful and versatile research laboratories that science has ever seen. Placed over the obscuring atmosphere, it will revolutionize astronomy by enabling clearer and more detailed observations of the solar system and beyond. Dark and uncharted parts of the sky will be studied, and mysterious objects, like quasars and black holes, will be better analyzed. Looking down on the panoramic view of earth, the space station will visually and electronically study environmental pollution, climatic global changes, and weather patterns. Exposure of living organisms to the low gravity and other unique surroundings of space will provide valuable information to medicine and the health sciences.

Since the birth of NASA in 1958, the United States has signed more than a thousand agreements with more than one hundred nations for cooperation in the space program. Several nations have provided NASA with sites for vital tracking and communications stations. NASA launch vehicles have carried satellites owned by other nations into orbit. Experiments built and owned by other nations or sometimes jointly owned with the United States have been flown on NASA spacecraft. In turn, American scientists have flown experiments in foreign spacecraft. Space research information has been shared extensively by all nations, and joint space ventures involving astronauts from other nations are planned.

This space station is composed of communications antennas, two solar arrays, a space radiator, rotating pallets capable of holding space science and applications payloads, an airlock joining two manned modules, and a logistics module.

An artist's concept of a space station. It consists of a number of modules clustered with large winglike solar panels. A space shuttle is shown docked at the station during a mission to resupply the station or rotate the crew.

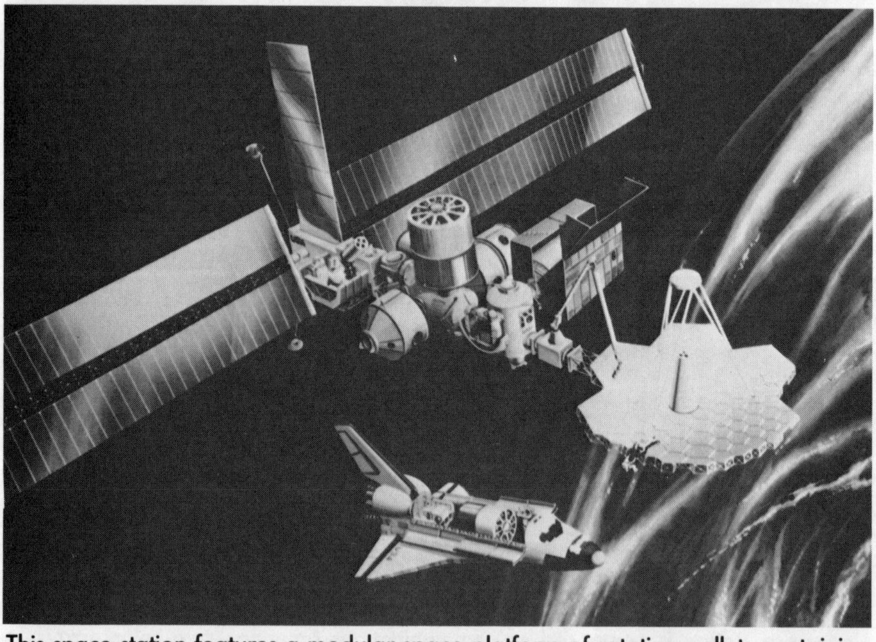

This space station features a modular space platform of rotating pallets containing space science and applications payloads with airlocks joining three manned modules. The space shuttle hovers near a large antenna platform.

Subscribing to the same spirit of global cooperation, the space station promises to offer an abundance of opportunities for experts from different nations to work side by side. When President Reagan directed NASA to build the space station in January 1984, he emphasized foreign contributions. He said, "We want our friends to help us meet these challenges and share in their benefits. NASA will invite other countries to participate so we can strengthen peace, build prosperity, and expand freedom for all who share our goals." Such an international perspective on science not only makes financial sense because of cost sharing, it also becomes a symbol of goodwill among nations that may one day help reduce world tension. On the other hand, unfortunately, the space station has the potential to become a superspy and battle station that could create fear and suspicion in some nations.

Colonies in Space

In 1772 the French mathematician Joseph Louis Lagrange (1736–1813) proved that between two large masses in space there exist regions where a small body will remain always at rest, experiencing a zero net gravitational field. The earth-moon system has two such points, named L4 and L5. These regions in space are ideal locations for building a permanent space colony. Such a colony, in the form of a giant space station, would have permanent stability without the need for periodic corrections.

Gerard K. O'Neil, a professor of physics at Princeton University, is a leading authority on colonies in space. Mainly through his efforts, conferences have been held at which scientists, engineers, lawyers, and social scientists have met on a number of occasions. These conferences dealt with the feasibility, benefits, and problems of space colonies, incorporating available data and substantial quantitative calculations. The startling conclusions of these conferences were not only that space colonies are possible but that they could solve problems of famine, pollution, energy, and living space even if the human population were to grow to twenty thousand times its present number.

O'Neil has predicted space colonies of ten thousand population inside a giant prefabricated cylinder, sphere, or doughnut-shaped wheel, rotating to provide artificial gravity and using sunlight for energy, heat, and light as well as for raising crops and food animals. They would include residences, schools, stores, hospitals, entertainment centers, and even parks and lakes. Using mirrors in giant greenhouses, these colonies would raise an abundance of crops that require either a continuous or a rhythmic day-night cycle. Power for spinning as well as for the daily necessities of the colony would be derived from turbogenerators driven by sunlight-produced steam.

It was suggested that, with food and energy in abundance and a controlled pleasant climate, life in these colonies would be more desirable than in most regions on earth. Recreation and sports would be available, and an abundance of free time would allow the development of art, philosophy, metaphysics, and a unique government system that fulfilled the needs of individual colonies. The options of creativity and individual expression would be unlimited.

A major problem in the construction of the permanent colony will be the manufacture and delivery of the raw materials. The possibility of sending these from earth is quickly discarded because of the high cost of launching them. An ideal alternative is the moon, which can provide as much as 98 percent of all required materials. Its low gravity and absence of frictional air would reduce shipping costs to a construction point such as L5 to one twentieth of what it would cost from earth.

The analysis of moon samples brought back by *Apollo* has shown them to be composed of 40 percent oxygen by weight, 20 percent silicon, and 20 to 30 percent metals. Many metals have been found on the moon, including aluminum and titanium, iron, and silicates for glass. Lunar soil could even be used as the bulk of the soil needed for agriculture in the colony. It would first have to be enriched with nitrates and trace elements from Earth. Of the elements most needed in the construction of a space colony, only carbon, nitrogen, and hydrogen are deficient in the lunar soil. These will have to be transported from earth.

Setting up a mining and industrial colony on the moon is therefore a necessary step toward the colonization of space. According

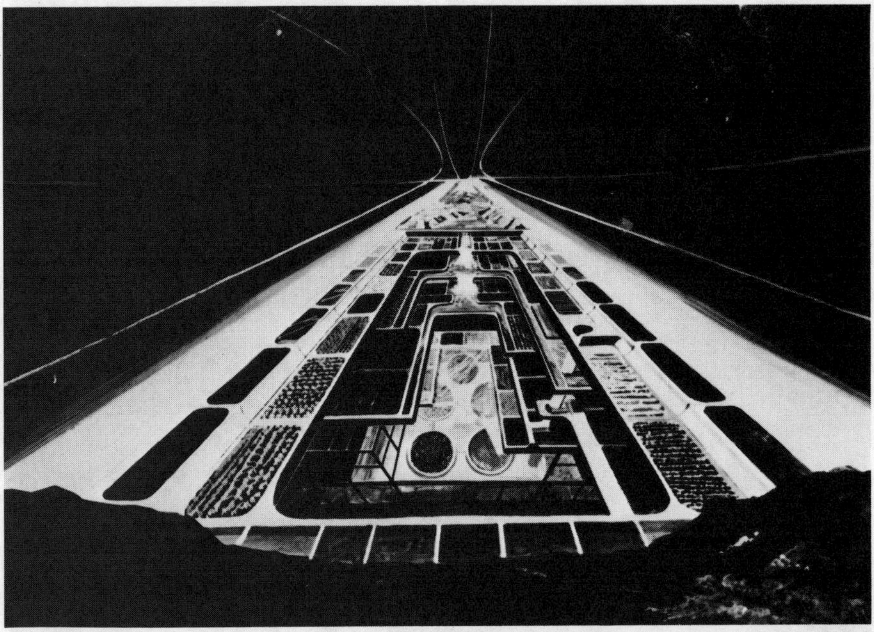

The agricultural area of a space colony. The half-mile-long farm can grow wheat and soybeans, and support rabbits and as many as 2,800 cows. The water tanks used for irrigation could hold about 260,000 fish. With controlled moisture, sunlight, and heat this space farm could yield far more than a comparable-size farm on earth.

to science writer Mitchell R. Sharpe, the colonizing of the moon will develop in three stages.[26] The first stage will be short-term exploration to select the best site, which will provide natural shelter and be rich in minerals as well as water. Prominent American and Soviet scientists believe that there are regions on the moon that are very rich in ice. Ice cannot, of course, be found on the surface because of the high temperatures and vacuum environment. It is, however, believed that large quantities of ice are trapped under a layer of lunar soil.

In the second stage, longer-term explorations will be undertaken, utilizing long-range surface vehicles, laboratories, and underground shelters. These shelters will be built easily, and they will provide an effective, low-cost, comfortable temperature environment. Temperatures on the surface of the moon vary from 120°C in daytime down to −115°C in the evening. Temperatures below two feet of lunar soil, which acts as an excellent insulator, remain constant all the time.

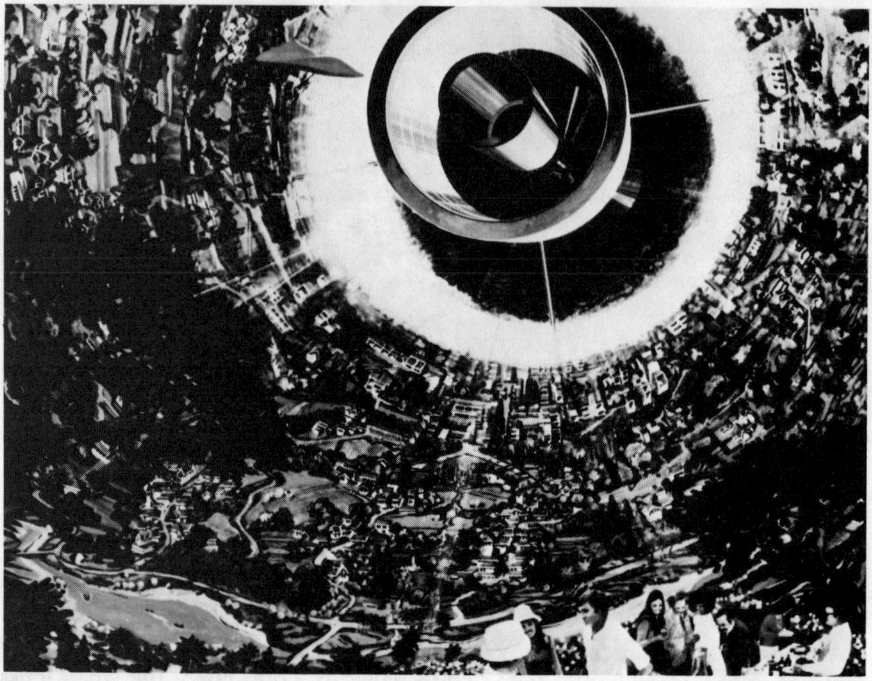

The interior of a ten-thousand-person space settlement. Natural sunshine is brought inside by external mirrors. Rotation of the sphere at 1.9 rpm would produce gravity like earth's at the "equator," but at the "poles" it would be zero. In these areas colonists could enjoy low-gravity sports. A corridor at the axis would permit floating in zero gravity out to exterior facilities such as observatories, docking ports, and industrial and agricultural areas.

The third and final stage will involve the establishment of a large-scale permanent self-sustaining colony. Shelters will be made using explosives to create cavities below the surface. These will then be "lined up" with an inflated balloon that will push against the walls. This will be necessary to preserve precious oxygen. Adjacent cavities will be connected with tunnels. Their roof will be a plastic dome with pressurized isolated airlock compartments for entry-exit traffic.

Under similar plastic dome structures, growing plants will consume carbon dioxide and produce oxygen and food. The production of oxygen and water will be the most critical life-support activities of the crew. Besides extracting water from ice deposits beneath the lunar surface, NASA scientists have managed to produce water from ilmenite, a lunar mineral that contains iron and titanium oxides. Heated by parabolic mirrors to 1,390°C, it is flashed

The interior of a ten-thousand-person space colony constructed almost entirely of ore mined from the moon.

with hydrogen, which combines with released oxygen to form steam. Without the hydrogen the method can also be used for the production of oxygen. Other heavy lunar industry will be the production of metals and glass. The smelting of rocks will be done by focusing large parabolic mirrors into a sealed furnace chamber, achieving temperatures of 2,250°C or more.

There will be many other benefits from lunar stations besides providing raw materials for the space colony. Geologists will carry out long-term investigations to answer many questions about the age and origins of the solar system. Lunar optical and radio telescopes, free from any atmospheric absorption, will provide a more accurate picture of the sky and a better understanding of the sun's activities. One practical use of such information will be to evaluate the earth's albedo, the ratio of incident to reflected sunlight, which governs the heat balance of the atmosphere and therefore the weather.

The moon can also be used as a spaceport for future launchings of manned or unmanned spacecraft to the planets or beyond. It

A small transportation vehicle waits for lift-off from a mining town on the moon. Lunar materials will be the prime source of metals and oxygen for a proposed space colony. To the right are the mines and the living quarters. To the left a magnetic track accelerates lunar materials and hurtles them into space, where they will be collected for construction of the proposed space colony.

can also provide the raw materials needed for the construction of another family of sky objects: orbiting power stations. Their giant "sails," with hundreds of square miles of surface area covered with solar cells, will convert solar energy into electricity twenty-four hours a day. Electricity converted into microwaves will then be beamed down to receiver stations for distribution. Feasibility studies show that such projects are both possible and very cost-effective once a lunar colony is already in place.

If space is to be considered as a giant continent, our trip to the moon was merely a touch on its shoreline. Space colonization no

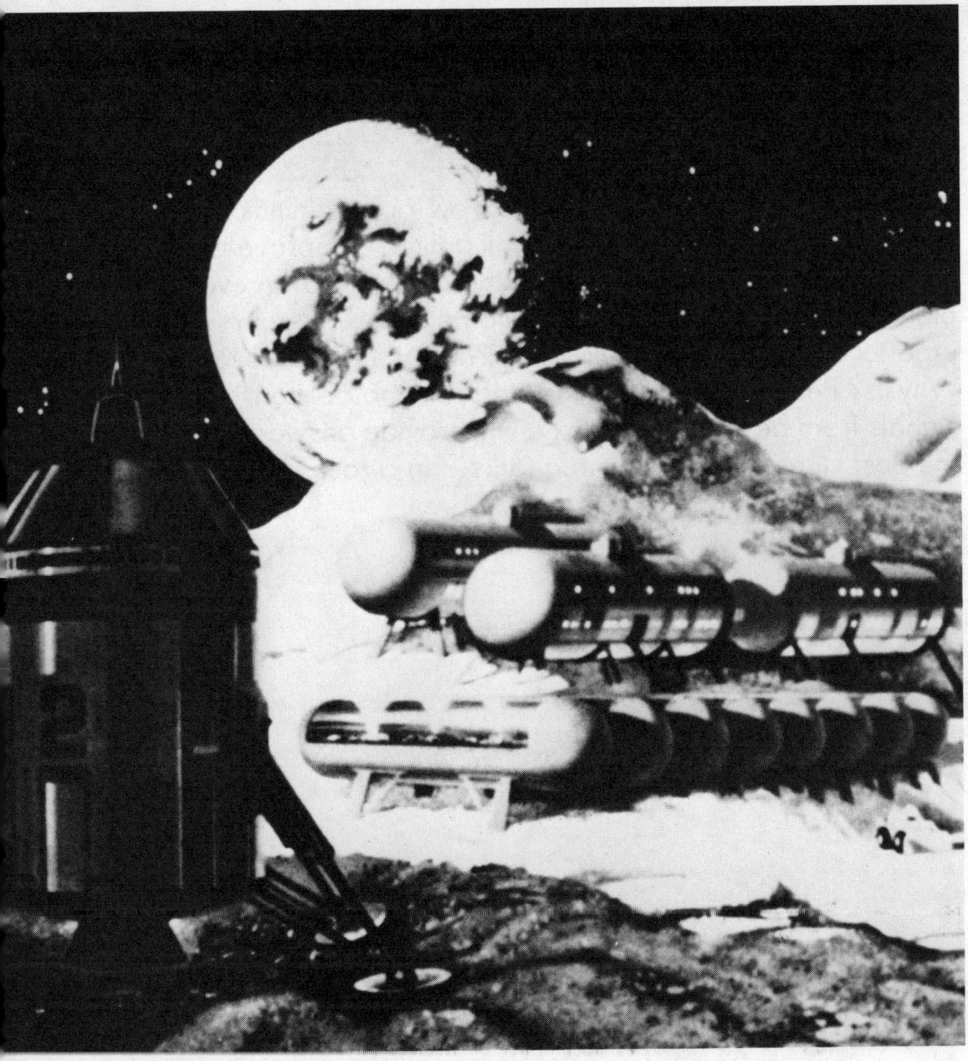

longer belongs in science fiction but has become an exciting reality. Our curiosity to know what's beyond has taken us up there and will continue to drive us into further exploration and colonization.

The new frontier of space will generate a great many opportunities for international cooperation that one day will hopefully lead to a strong global identity. The need to militarize space, which at the moment some see as essential for national defense, may then no longer be necessary. The economic and technological opportunities of space will also come to the rescue of some critical needs like energy and raw materials and reduce the cost and improve the quality of others.

With new discoveries will come new opportunities as well as challenges. Treating both wisely will definitely improve the human condition and elevate our role in the universe to an even higher plane. One day it may all make some sense. Why this powerful urge to explore and to colonize? What is the purpose of it all? Why us? Perhaps the beckoning stars, the most distant, mysterious signals from beyond our galaxy, will provide answers. Or maybe the best clues will come from quarks... in a rose or an ice cream cone, if not directly from a star.

Notes

1. I. I. Rabi in F. J. Rutherford, G. Holton, and F. G. Watson, *The Project Physics*, Holt, Rinehart & Winston, 1973, viii.

2. Lucretius in R. C. Trevelyan, *Transactions from Lucretius*, George Allen & Unwin Ltd., 1920.

3. Robert Boyle in Joan Solomon, *The Structure of Matter*, David and Charles Ltd., 1973.

4. J. J. Thomson, *Mosaic* (Journal of National Science Foundation), vol. 10, no. 5 (September 1979), 4.

5. Ernest Rutherford in F. J. Rutherford, G. Holton, and F. G. Watson, *The Project Physics*, Holt, Rinehart & Winston, 1973.

6. Neils Bohr in J. Bronowski, *The Ascent of Man*, BBC Publications, 1973, 256.

7. Victor Weisskopf, *Mosaic* (Journal of National Science Foundation), vol. 10, no. 5 (September 1979), with permission.

8. Enrico Fermi, *Mosaic* (Journal of National Science Foundation), vol. 10, no. 5 (September 1979), 6.

9. William Blake, "Auguries of Innocence," in Jacob Bronowski, *The Ascent of Man*, BBC Publications, 1973.

10. Guglielmo Marconi in Ralph Stein, *The Great Inventions*, a Ridge Press book, Playboy Press, 1976, 117.

11. Walter H. Brattain in Belle Melville Grosvenor (ed.), *Those Inventive Americans*, National Geographic Society, 1971, 216.

12. Albert Einstein in Banesh Hoffman, *Relativity and Its Roots*, Scientific American Books, 1983, 129.

13. Albert Einstein in J. M. Pasachoff and M. L. Kutner, *Invitation to Physics*, Norton, 1981, 160.

14. Albert Einstein in J. M. Pasachoff and M. L. Kutner, *Invitation to Physics*, Norton, 1981, 161.

15. Philipp Lenard in Philipp Frank, *Einstein: His Life and Times*, Alfred A. Knopf, 1963, 253.

16. Anonymous in Philipp Frank, *Einstein: His Life and Times*, Alfred A. Knopf, 1963, 253.

17. Anonymous in Philipp Frank, *Einstein: His Life and Times*, Alfred A. Knopf, 1963, 255–256.

18. Albert Einstein, *Ideas and Opinions*, Bonanza Books, 1954, 9–10.

19. Albert Einstein, *Ideas and Opinions*, Bonanza Books, 1954, 11.

20. Albert Einstein in Timothy Ferris, *Science 83*, 38.

21. Albert Einstein, *Ideas and Opinions*, Bonanza Books, 1954, 39.

22. Albert Einstein in J. Bemporad, A. Segal, J. D. Spiro, and R. S. Widom (eds.), *Focus on Judaism, Science and Technology*, UAHC, 1970, 23.

23. Albert Einstein in the brochure celebrating Einstein's centennial, *1879–1979: A Centennial Exhibit*, American Institute of Physics, 1979.

24. Metrodoros of Chios in L. Berman and J. C. Evans, *Exploring the Cosmos*, Little, Brown & Co., Inc., 1977, 376.

25. Leonardo da Vinci in A. C. Clarke and editors of *Life Magazine, Man and Space*, Life Science Library, Time Incorporated, 1964, 21.

26. Mitchell R. Sharpe in the *1972 Britannica Yearbook of Science and the Future*, 11.

Suggested Reading

H. Allaway. *The Space Shuttle at Work.* NASA, 1979.
A. D. Anderton. *Spacestation.* NASA, 1985.
F. J. Baugher. *On Civilized Stars.* Prentice Hall Press, 1985.
Russell Bourne (ed.). *The Smithsonian Book of Invention.* Smithsonian Exposition Books, 1978.
Jacob Bronowski. *The Ascent of Man.* BBC Publications, 1973.
Nigel Calder. *Einstein's Universe.* Greenwich House, 1979.
J. D. Crump (ed.). *Frontiers of Science.* National Geographic Society, 1982.
Albert Einstein. *Ideas and Opinions.* Dell, 1976.
L. C. Epstein. *Relativity Visualized.* Insight Press, 1985.
W. Froehlich. *Spacestation: The Next Logical Step.* NASA, 1985.
George Gamow. *Mr. Tompkins in Paperback.* Cambridge University Press, 1981. (Nontechnical description of relativity, quantum theory, and nuclear structure.)
Belle Melville Grosvenor (ed.). *Those Inventive Americans.* National Geographic Society, 1971.
Jeff Hecht and Dick Teresi. *Laser: Supertool of the 1980s.* Ticknor and Fields, 1982.
J. H. Mulvey (ed.). *The Nature of Matter.* Clarendon Press, 1981.
NASA. *NASA: The First 25 Years: 1958–1983.* NASA, 1983.
K. Gerard O'Neil. "Colonies in Space," *1977 Yearbook of Science and the Future.* Britannica.
Carl Sagan. *Cosmos.* Random House, 1980.
E. G. Tauber. *Man's View of the Universe.* Crown, 1979.
C. A. Walker et al. (eds.). *Too Hot to Handle.* Yale University Press, 1983. (The science and politics of nuclear waste management.)

WGBH, Boston. *NOVA: Adventures in Science.* Addison-Wesley, 1982.

Gary Zukav. *The Dancing WuLi Masters.* William Morrow, 1979. (A nontechnical overview of the new physics.)

Relevant articles in the *Journal of the National Geographic Society*:

"The Chip," vol. 162, no. 4 (October 1982).
"The Universe," vol. 163, no. 6 (June 1983).
"Satellites," vol. 164, no. 3 (September 1983).
"The Laser," vol. 165, no. 3 (March 1984).
"Genetic Engineering," vol. 166, no. 6 (December 1984).
"The Planets," vol. 167, no. 1 (January 1985).
"The Atom," vol. 167, no. 5 (May 1985).

Index

120